DIY Lithium Batteries
How to Build Your Own Battery Packs

Written by
Micah Toll

Printed in the United States of America
Toll Publications, 2017
ISBN 978-0-9899067-0-8

To my wife Sapir, who tolerates me more than anybody should have to…

Table of Contents

Safety disclaimer:

Lithium batteries can be dangerous. They contain a large amount of energy in a small volume, and are specifically designed to release that energy quickly. When used properly, they can be a safe and efficient way to power nearly anything. However, when used improperly or carelessly, lithium batteries have the potential to cause devastating fires that can and have resulted in loss of property and life.

This book is meant to be an educational guide. Please do not attempt to recreate anything you see in this book or on the internet without professional guidance and training. Always use proper safety equipment. Always engage in property safety practices. Never leave lithium batteries charging unattended. Always use your head, be smart and be safe.

Chapter 1: Introduction

Lithium batteries have existed in various forms since the 1970's, and innovations in the 80's and 90's have led to the familiar lithium battery cells that we know today. Current research on lithium batteries has produced battery cells capable of extreme performance, for example, 100% recharging in just a few seconds. However, these current advances are strictly experimental and won't see commercialization for many years, potentially decades. The information in this book covers the types of lithium batteries that are commercially available today and will likely remain available well into the future.

Uses for lithium batteries

Today, lithium batteries are used for a seemingly endless number of applications. They can be found everywhere from electric vehicles to NASA's spacesuits. Due to their lightweight and energy dense properties, lithium batteries are perfect for an incredibly wide range of applications.

In the past, lithium batteries were mostly used by original equipment manufacturers (OEMs) for use in consumer products. These big manufacturers built lithium batteries suited to their needs for specific products or large clients. If a hobbyist wanted a battery size or shape that didn't exist, he or she was out of luck. However, today there are many lithium batteries and cells that are readily available directly to consumers for use in, well, whatever we want!

I was introduced to the world of custom lithium batteries during my time spent working in the do-it-yourself (DIY) electric bicycle industry. I have been building lithium batteries for electric bicycles for years, largely because the number and variety of lithium batteries available to the consumer market has always been frustratingly small. If I wanted a specific size battery pack but it didn't already exist, I had no other choice than to make it myself. This opened up a whole new world to me. Suddenly, I could build batteries of any voltage, any capacity and most importantly, any size and shape that I wanted.

But DIY lithium batteries aren't only limited to the world of electric bicycles. There are thousands of applications for DIY lithium batteries!

Even though electric cars are becoming increasingly available in the consumer market, it can still be cheaper (and more fun) to build your own. Many people convert all kinds of vehicles into electric vehicles, and they need batteries to do it. Unless you want to buy an expensive, purpose built electric car battery, you'll need to know how to assemble your own large battery pack from lithium battery cells.

Just like electric vehicles, home batteries are also becoming increasingly popular. A lithium battery in the back of your closet or hidden in your garage can power a house for days in the event of a power outage. They are also great for storing energy that has been generated on site, such as from solar panels or wind turbines. Home battery storage systems like the Tesla Powerwall are great OEM products, but you can still build your own custom system suited for your unique needs. All you need is the battery know-how!

Drones, wearables, backup batteries, toys, robotics and countless other applications are all ripe for custom DIY lithium batteries. This book will teach you how to design and build lithium batteries for all of these uses and more. Prepare yourself, because by the time you finish this book, you are going to be full of knowledge and rearing to go out and power the world!

How lithium battery cells work

Despite undergoing years of research and development, the electrical and chemical processes that allow lithium battery cells to function is actually fairly simple. As lithium ion batteries are by far the most common form of lithium battery cells, we'll take a look at how a typical cell works here.

A lithium-ion cell is composed of four main parts:

- Cathode (or positive terminal)
- Anode (or negative terminal)
- Electrolyte
- Porous separator

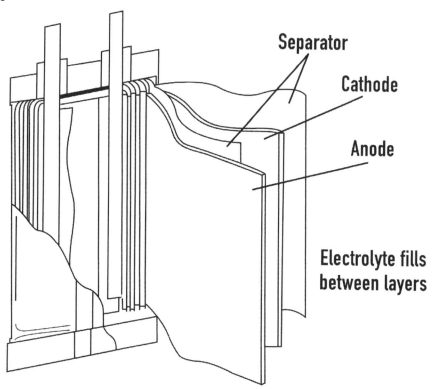

The cathode varies between different types of cells but is always a lithium compound mixed with other materials. The anode is almost always graphite, and sometimes includes trace amounts of other elements. The electrolyte is generally an organic compound containing lithium salts to transfer lithium ions. The porous separator allows lithium ions to pass through itself while still separating the anode and cathode within the cell.

When the cell is discharged, lithium ions move from the anode to the cathode by passing through the electrolyte. This discharges electrons on the anode side, powering the circuit and ultimately any device connected to the circuit. This process is demonstrated in the diagram below. When the cell is recharged, this process is reversed and the lithium ions pass back from the cathode to the anode, which is opposite to the diagram below.

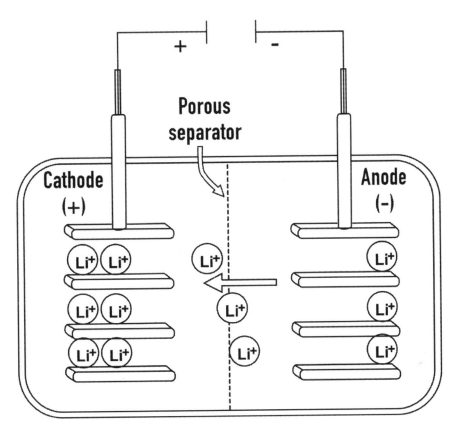

The actual process is quite simple. The major differences (and where things get more complicated) are in the shape of the cells and their slight chemical changes. We'll cover all of that information in the next two chapters.

Chapter 2: Form factors of lithium cells

Lithium battery cells are available in a number of different form factors, yet their underlying construction is always the same. All lithium battery cells have a positive electrode (cathode), a negative electrode (anode), an electrolyte material and some type of porous separator in between that allows lithium ions to move between the cathode and the anode. We'll talk about how changes in the chemistry of different li-ion cells can affect them in the next chapter. For now, the main difference between various shapes of lithium cells is the way they are assembled.

Pouch cells

Pouch cells are the simplest form of lithium battery cells. They look like a tin foil bag (or pouch, get it?) and have two terminals at an edge of the pouch. Inside the pouch is a cathode and anode on opposite sides separated by the porous separator and with the electrolyte on either side of the separator. This cathode-electrolyte-anode sandwich is folded back and forth many times within the pouch to increase the capacity of the battery.

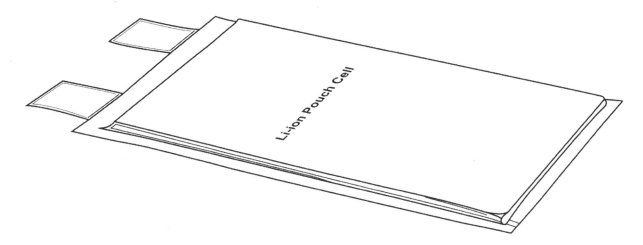

There are no standard sizes for pouch cells. They are produced by many different companies and are often designed to exact sizes for specific products, such as cell phones, to ensure that they take advantage of the maximum possible usable space. Production at such high volumes allows for the lack of standardization of sizes. When you can afford to have a million battery cells made, it becomes less important if a dealer has your size in stock or not.

The advantage of pouch cells is that they are lightweight and cheap to produce. The main disadvantage is that they have no exterior protection and thus can be damaged if they aren't enclosed in some form of protective case. A lack of a hard exterior case means they are the lightest and most space efficient way to produce a lithium battery cell. Pouch cells are often used in consumer devices such as laptops and cellphones due to their efficient use of space. These devices also serve as protection for the fragile pouch cell inside.

Pouch cells actually perform better when they are contained in some time of rigid or semi-rigid structure that can apply a slight amount of pressure to the cells. This helps keep all of the layers of the cells in close contact and prevents micro-delamination which can degrade cell performance over time.

When a pouch cell ages it can begin to expand or "puff" as it is sometimes called in the industry. This is often due to small interior shorts that occur over time as the battery ages, creating gas that puffs up the cell. Because pouch cells are entirely sealed, the gas has nowhere to escape and thus creates the puffy, pillow-like appearance.

The expansion of the pouch cell results in a reduction in performance of the cell as the layers of the cell further delaminate. Some degree of gas buildup can be retained by the pouch structure, but when the gas buildup becomes too great, the pouch can rupture explosively. This is a rare yet well documented phenomenon. The rupture releases a large amount of flammable gas - not a great situation to be in.

Prismatic cells

Prismatic cells are quite similar to pouch cells, except that they have the addition of a rigid rectangular case outside of the cell. This gives the cell a rectangular prism (or prismatic) shape. Prismatic cells are therefore slightly less space efficient than pouch cells, but are also more durable than pouch cells. While pouch cells must be handled carefully, prismatic cells can withstand more jarring, though they can still be fragile.

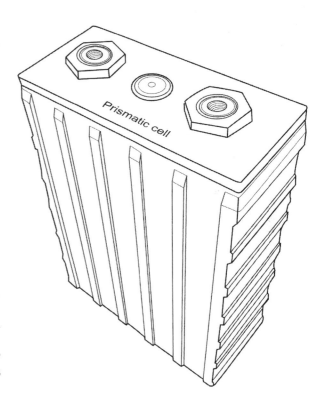

Unlike pouch cells that have thin tab terminals, prismatic cells often have threaded terminals that allow a nut or bolt to be used for connections. This makes it easier to join prismatic cells into larger battery modules. Large prismatic cells of 20 Ah to 100 Ah or more are often used in very large energy storage devices such as home batteries or DIY electric vehicles. There aren't standard dimensions for prismatic cells, but they often come in various capacities with 5 - 10 Ah increments.

Cylindrical cells

Cylindrical cells are the AA-style batteries that we are all familiar with from remote controls, flashlights and other consumer electronics. They come in a variety of sizes (most are larger than standard AA batteries) but all share the same cylindrical shape and rigid metal case.

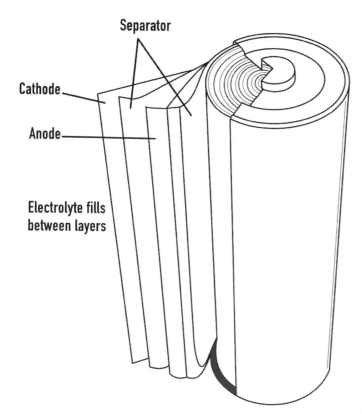

Separator

Cathode

Anode

Electrolyte fills
between layers

Cylindrical cells are produced by rolling up what amounts to the same contents of a pouch cell, then placing it inside of a metallic cylinder with a positive and negative terminal at either end of the cylinder. These cells are not as space efficient due to the rolling of the inner layers and the addition of the cylinder wall and end caps. However, cylindrical cells are the most robust type of lithium battery cell and don't require any external frame or support.

Unlike pouch cells and prismatic cells, cylindrical cells are produced in standard sizes. The most common lithium battery cylindrical cell is the 18650 cell, named for its 18 mm diameter and 65 mm length. The 18650 is the cylindrical cell most commonly used in laptops, power tools, flashlights and other devices that require cylindrical lithium cells. Two other common sizes of cylindrical cells are the 14500, which is 14 mm in diameter and 50 mm in length and is the same size as a standard AA battery, as well as the 26650, which is 26 mm in diameter and 65 mm in length. The 18650, which falls right in the middle of the three most common cylindrical standard sizes, has seen the most widespread use and is available from the highest number of manufacturers.

In 2017, Tesla began producing the new 21700 cell format that they co-developed with Panasonic. The 21700 is a slightly larger cell than the 18650 and comes with a decent boost in capacity compared to current 18650s. The cell was designed specifically for Tesla's vehicles and so it will likely take a few years until it becomes available for DIYers to incorporate into their own battery projects.

There are also a series of LiFePO4 cylindrical cells made by the company Headway that are available in the 38120 and 40152 sizes, which are 38mm in diameter, 120 mm in length and 40 mm in diameter, 152mm in length, respectively. These are obviously much bigger cylindrical cells and have much higher capacities than 18650 cells. These cells are the only cylindrical cells that have bolted terminals for easy connections Most other cylindrical cells must be spot welded to connect them together.

Chapter 3: Types of lithium cells

All lithium battery cells aren't created equally. There are a few different chemistries of lithium batteries that have very different properties and specifications. They all have their own unique advantages and disadvantages, so let's compare them here.

Lithium ion (Li-ion)

Li-ion is the most common type of lithium battery used in consumer electronics like cellphones, laptops, power tools, etc. They have the highest energy to weight ratio and are also some of the most energy dense cells, meaning you can pack a lot of energy into a small volume.

Depending on the exact type, li-ion cells are relatively safe cells, at least as far as lithium batteries go. Most li-ion cells won't just burst into fire if they are punctured or the cell is otherwise heavily damaged, though this can happen with some types of li-ion and has been observed many times. The chance of fire is always present in lithium batteries, but is usually caused by negligence or abuse of a lithium cell or battery. Short circuiting a battery is one common example of such negligence, but we'll talk more about short circuits in Chapter 7 on safety.

Li-ion cells also have relatively long cycle lives. The shortest are rated for around 300 cycles until they reach 70-80% of their initial charge capacity, while the longest can last for over 1,000 cycles. There are of course ways to stretch the number of cycles that you can get out of a lithium cell even further, which we'll talk about in Chapter 14. Just based on manufacturer's ratings though, li-ion cells are middle of the road for cycle life, as compared to the other two major chemistries that we'll talk about next.

Cost is always an important factor when choosing components for any project. Li-ion cells fall in the middle range of lithium cell prices (you might be noticing that li-ion is something of the "Goldilocks" chemistry - it's right in the middle on many of these specifications). There are cheaper chemistries (RC lipo) and more expensive chemistries (lithium iron phosphate), which leaves standard li-ion somewhere in the middle in terms of price.

Where li-ion really shines is in availability. Because this is the most widely used lithium battery chemistry, it's also the most widely available in different sizes, shapes, capacities and slight chemical variations that have different effects on the performance.

One of the most common and easiest to work with formats of li-ion cells is the 18650 cylindrical cell that we talked about in the previous chapter. There are dozens and dozens of great quality, top brand 18650 li-ion cells, plus hundreds of other off-brand and generic 18650 li-ion cells as well. Because 18650s are so commonly used in OEM products including everything from electric vehicles to power tools, they have been developed with a wide range of specifications. You can find cheap, low power 18650 li-ion cells like Samsung ICR18650-26F cells that are perfect for simple, low power projects, or you can find insanely powerful Sony US18650VTC5, which have the same approximate capacity, size and weight but can provide over 600% more power!

Anyone who makes use of high power 18650 cells owes a big debt of gratitude to the electric power tool industry, by the way. They were some of the first to demand higher power cylindrical li-ion cells, which spurred the battery industry to respond and meet that demand with new and ever higher power cells. Thanks to power drills, you can now find li-ion cells that contain a massive amount power in something the size of your thumb.

It's difficult to say which projects are best suited for li-ion use, mostly because different li-ion cells span such a large range of specifications and properties. However, if your project has space and weight limitations as well as moderate to high power needs, li-ion is likely a good option for you.

Most li-ion cells have a nominal voltage of between 3.6 V to 3.7 V and are usually rated for a discharge-charge voltage range of 2.5 V - 4.2 V. Li-ion cells are usually rated for maximum capacity at this voltage range (i.e. charging to 4.2 V, then discharging down to 2.5 V) but it is recommended to avoid draining li-ion cells all the way to 2.5 V very often. They can handle it, but it reduces their expected lifetime. Most battery management systems (BMSs) for li-ion batteries cut off discharge at around 2.7 V - 2.9 V per cell. Discharging below 2.5 V will cause irreparable damage to the cell, resulting in the cell not holding its rated capacity or sustaining its rated discharge current.

Some newer li-ion chemistries are becoming commercially available that are designed to be charged as high as 4.3 V - 4.4 V. These are still the exception, and most li-ion cells should never be charged to higher than 4.2 V. Always check the manufacturer's recommendations for highest rated charging voltage. Overcharging a li-ion cell not only reduces its lifetime, but can be dangerous as well.

There are a number of unique li-ion chemistries that are all contained within the larger class of li-ion cells. They all share very similar or identical anode (negative terminal) materials but have unique cathode (positive terminal) materials. The different li-ion chemistries are listed below.

Lithium Manganese Oxide ($LiMn_2O_4$ or li-manganese)

$LiMn_2O_4$ gets its name from the use of a manganese matrix structure in the cathode. It was developed in the late 1970's and early 1980's, making it one of the first commercial li-ion chemistries. $LiMn_2O_4$ can handle relatively high power in very short bursts and offers high thermal stability. This makes it one of the safer li-ion chemistries because higher temperatures are required to cause thermal runaway.

$LiMn_2O_4$ cells can also be tweaked for either higher power or higher capacity at the expense of each other. The downside to $LiMn_2O_4$ is its relatively lower cycle life compared to other li-ion chemistries. An example of a $LiMn_2O_4$ cell is the LG 18650 HB2.

Lithium Cobalt Oxide (LiCoO$_2$ or li-cobalt)

LiCoO$_2$ was developed around the same time as LiMn$_2$O$_4$ and was also one of the earliest forms of commercially available li-ion cells. It uses a layered cobalt structure in its cathode. LiCoO$_2$ is known for its relatively low cost and high capacity, but generally has a lower current rating and only moderate cycle life. It also has a lower thermal runaway temperature, making it somewhat less safe than other li-ion chemistries. An example of LiCoO$_2$ is the Samsung ICR18650-26F cell.

LiCoO$_2$ is also the basis for the much more dangerous RC lipo batteries that we'll discuss later in this chapter. In RC lipo batteries, the chemistry is altered to produce a much more powerful cell capable of sustaining extremely high discharge current. This increased power comes at the expense of safety, weight and cycle life.

Lithium Nickel Manganese Cobalt Oxide (LiNiMnCoO$_2$ or NMC)

LiNiMnCoO$_2$ is a fairly new chemistry that is still undergoing constant development. NMC falls in the sweet spot of improving upon the drawbacks of many previous types of li-ion chemistries while retaining their benefits. NMC shares many of the advantages of both LiCoO$_2$ and LiMn$_2$O$_4$ chemistries. By combining cobalt and manganese, then including nickel, NMC cells have demonstrated relatively high power, capacity, and safety.

By adjusting the ratio of cobalt, manganese and nickel in the cathode as well as including other trace elements in both the cathode and anode, NMC cells can be tweaked for improved performance in nearly any measurement category. Other chemistries are capable of achieving better performance in some categories, but NMC have some of the highest all-around performance figures of any lithium battery chemistry.

This makes NMS an excellent "all around" chemistry. It doesn't have the highest performance in any single category, but it has some of the highest average performance of any chemistry. An example of an NMC cell is the Samsung INR18650-25R, which is optimized for relatively high power and medium capacity.

Lithium Nickel Cobalt Aluminum Oxide (LiNiCoAlO$_2$, NCA, or NCR)

LiNiCoAlO$_2$ is very similar to the NMC chemistry above, but with aluminum swapped for manganese in the cathode. The addition of aluminum helps NCA cells achieve the highest capacity of all li-ion chemistries. The downsides are a slight decrease in cycle life and power as compared to most other chemistries. An example of an NCA cell is the Panasonic NCR18650B cell, which was used for most or all of Tesla's early electric vehicles.

Like NMC, NCA is a very promising chemistry for future development of li-ion cells. It is best suited for high capacity, energy dense purposes. This is why it was selected by Tesla for use in their electric cars. NCA excels in packing the most energy into the smallest space. With a large enough battery, its lower relative power can be mitigated. However, continued research and incremental improvements are helping to increase the power of this type of cell, making it quite competitive with NMC.

Lithium polymer (li-poly or lipo or RC lipo)

There's an immense amount of confusion out there regarding lithium polymer batteries. This is mostly because the cells that the term was originally created for and the cells that most people today call lithium polymer cells are not the same thing.

Remember when we discussed how lithium cells are made? How they have an anode, a cathode, and a liquid or more commonly gelled electrolyte material in between the two? Right. So the original term "lithium polymer battery" referred to a new type of lithium cells that used a solid (sometimes referred to as "dry") electrolyte instead of the common liquid or gelled electrolyte. The solid electrolyte used in these experimental cells was a polymer, or plastic material, giving rise to the name "lithium polymer battery".

This new technology for dry electrolyte batteries promised incredibly safe batteries. However, it never made it out of the laboratory on any large scale. The problem was that the dry electrolyte didn't conduct electricity very fast at ambient temperature. That meant that the batteries had to be constantly heated to function properly. This was obviously a deal breaker for most applications. Who wants a big heater built into their cell phone or laptop?

So the original lithium polymer batteries never really went anywhere. The problem with the name began when some manufacturers started referring to other cells that had a polymer packaging, namely pouch cells, as "lithium polymer" cells. This became confusing, as these cells didn't really have polymer electrolytes, but instead had their liquid electrolytes gelled with the use of an external polymer. These should really be called "lithium-ion polymer" cells to distinguish them from the original, non commercialized "lithium polymer" cells. However, once people started calling them lithium polymer cells, the confusion began.

But the confusion doesn't stop there! Because what people began calling lithium polymer batteries (which were actually lithium-ion polymer batteries in pouch cell formats) are actually nearly identical to the standard lithium-ion batteries that already existed. They have the same or similar cathode and anode materials and similar amounts of electrolyte. The main difference is that lithium-ion polymer batteries use a micro-porous electrolyte instead of the normal porous separator layer placed in the electrolyte of li-ion cells.

That means that all "lithium-ion polymer" and "lithium-ion" batteries available today are technically li-ion batteries. They're all similar and they all function by transporting lithium ions back and forth through an electrolyte. But the term "lipo", which is short for lithium polymer, has now commonly been used to refer to the shape and style of cells, namely pouch cells whose pouches are technically a polymer material. Because this use of the term "lipo" took off, many people now think that a lipo cell is another name for a pouch cell. In fact, a pouch cell is simply a type of battery cell structure and it can be used to make li-ion, $LiFePO_4$ or potentially other new chemistries in the future. So "pouch cell" describes the shape, not the chemistry. But now everyone seems to be running around referring to pouch cells as "lipos".

Lastly, there is an entire class of li-ion cells used for radio controlled (RC) toys and vehicles that are generally referred to as lipo batteries. These are extremely high power li-ion cells that are specifically used in the RC industry for their ability to provide the highest possible current.

The most common usage of the term lipo nowadays is to refer to these RC batteries. For that reason, in the remainder of this book, I will refer to these high power RC batteries as "RC lipo" batteries. This is not the original historical use of the term lithium polymer, but it is the commonly used convention today and thus it is how I will use it. When in Rome...

But please be aware that there is much confusion in the industry regarding the terms lithium polymer, li-ion and lipo. For our purposes, "RC lipo" will refer to li-ion batteries specifically designed for RC purposes, and all other lithium ion batteries will be referred to as li-ion. I won't use the term lithium polymer as any cell on the market being called "lithium polymer" today is really just lithium ion, and the real "lithium polymer" cells never really made it out of the lab.

Phew! Ok, I'm sorry that took so long, but it's important to point out the confusion and try to make sense of it. Now let's move on to actually learning about RC lipo batteries. Which should be fun, because these are the ones that go kaboom when you mess up with them.

Let's get this out of the way immediately: RC lipo batteries are the dangerous ones. These are the ones that are itching to burn your house down if you don't follow proper charging and discharging procedures. They can be safe, but they are also incredibly volatile when used improperly.

Alright, now that we've got that out of the way, let's look at what makes RC lipo batteries special. RC lipo cells are a specialty chemistry based on lithium cobalt that is suited for high power applications. They can provide super high discharge rates for long periods of time, and insanely high discharge rates for short periods (before they overheat and we get back to that unfortunate fire scenario I warned you about).

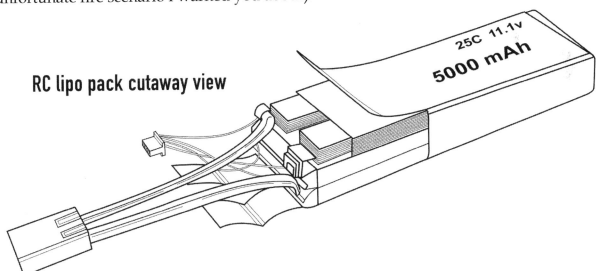

RC lipo cells are almost exclusively used in the remote control vehicle industry for applications such as RC drones, helicopters, planes, cars, etc. These devices require very high discharge rates from a small and lightweight battery. RC lipo cells aren't the lightest cells (those are variants of conventional li-ion cells), but they can provide much higher power for only a slightly higher weight.

RC lipo cells are also the cheapest lithium cells available. They cost much less than li-ion and LiFePO$_4$ cells (we'll learn about LiFePO$_4$ shortly), making them attractive for other applications that make use of DIY batteries, such as electric bicycles.

One major drawback (besides the fact that RC lipo are basically little bombs that can also power electronics) is that RC lipo cells have very short cycle lives. Reaching 200 cycles on a RC lipo cell would be considered fairly good performance. Some RC lipos can be pushed closer to 300 cells, but don't last as long as li-ion cells and can't even come near LiFePO$_4$ cells. (Note: these cycle counts are based on complete charging and discharging cycles. We'll talk about how partial charging and discharging can extend the life of nearly all types of lithium battery cells.)

Another issue with RC lipo cells is their more complicated charging process. While li-ion and LiFePO$_4$ cells are pretty easy to charge, especially when using a battery management system (BMS), RC lipo cells require more expensive balance chargers to ensure that all cells in a battery are maintained at the proper voltage and balanced with one another.

The reason for this is that when RC lipo batteries stray from their rated voltage range, they become incredibly volatile. Do a quick search on YouTube for "overcharging RC lipo cell" to get an idea of what I mean. It is critically important that RC lipo cells are charged within their specified voltage range. They should also never be discharged too low. Discharging a RC lipo cell below 2.5 volts and then charging the cell can result in combustion of the cell, especially at higher charging currents. For this reason, RC lipo cells must be monitored carefully during discharge as well to ensure that they are never drained too far.

It is possible to recharge RC lipo cells that have been over discharged, but it must be done at very low currents and can easily result in fire, depending how damaged the battery cell was. Ideally this wouldn't be attempted, but if it was, it should be done in a monitored environment and away from anything flammable.

RC lipo cells are electro-chemically similar to li-ion cells, and have a nominal voltage of 3.7 V. However, because care must be taken not to over-discharge the cells, it is not recommended to discharge them lower than 3.0 V. Aiming for a higher voltage of 3.2 V is considered safer. In the RC aircraft field, many pilots will stop flying when a battery reaches as high as 3.5 V, thus maintaining a larger safety margin. The maximum voltage of RC lipo cells should never exceed 4.2 V.

It should also be noted that these voltages are considered the "under load" voltage. Depending on the current load, a lithium battery cell (of any chemistry) will see a drop in voltage. This drop in voltage is known as voltage sag. An RC lipo cell should never drop below 3.0V while in use. If discharging stops at 3.0 V under load, the voltage when measured after the load is removed will return to a higher voltage, likely in the 3.3 V - 3.5 V range, though an even safer level for at rest voltage is around 3.7 V. For this reason, it is critically important to monitor RC lipo cells under load to ensure they never over-discharge beyond a safe limit.

Lithium Iron Phosphate (LiFePO₄)

Lithium iron phosphate is technically a subset of the more general li-ion class, but it is unique enough that it is often listed separately. LiFePO₄ cells are both heavier and less energy dense than most li-ion cells. This means that battery packs built from LiFePO₄ cells will be bulkier and more massive than li-ion or RC lipo batteries of the same voltage and capacity. The exact amount varies depending on the cell format, but you can expect a LiFePO₄ battery to be around twice as large and twice as heavy as a comparable li-ion battery.

LiFePO₄ cells are also some of the most expensive cells. Their cost varies based on many factors including cell size, format, vendor and location, but you can expect to pay around 20% more for LiFePO₄ cells than for li-ion cells of the same capacity.

Most commonly available LiFePO₄ cells also have a lower discharge rate, meaning they can't provide as much power, though this isn't always the case. Some cells, such as those made by the high quality battery company A123, can provide high power levels but cost a premium and are hard to source. Those are mostly sold to OEMs for use in consumer products like power tools.

It is common to hear LiFePO₄ being touted for its high discharge rate, but unless you source LiFePO₄ cells that are specifically designed for high discharge, most LiFePO₄ cells have relatively low discharge rates.

With all of these downsides, why would someone want to use LiFePO₄ cells? There are actually two big advantages for using LiFePO₄ - cycle life and safety. LiFePO₄ have the longest rated cycle life of all commonly available lithium battery cells. They are often rated for over 2,000 cycles. They are also the safest lithium battery chemistry available. While fires from LiFePO₄ cells have been documented, they are incredibly rare. The electrolyte used in LiFePO₄ cells simply can't oxidize quickly enough to combust efficiently and requires exceedingly high temperatures for thermal runaway, often higher than the combustion temperature of many materials.

So when should you use LiFePO₄ cells? The best applications for LiFePO₄ cells are projects that require long cycle lives and high safety, don't have critical space or weight limitations, and don't require very high levels of power (unless you specifically source high power LiFePO₄ cells).

LiFePO₄ cells have a nominal voltage of 3.2 V per cell and a discharge-charge voltage range of 2.5 V - 3.65 V. Just like with li-ion cells, discharging below 2.5 V will cause irreparable damage to the cell, though it isn't necessarily dangerous, like in the RC lipo cells that we just learned about.

General lithium battery cell summary

All of the information I have described for the lithium chemistries above is generally true for the majority of lithium battery cells that are commercially available today. That being said, there are exceptions in all of these cases. The newest technology for lithium battery cells far outpaces what you can currently buy. Right now there are lithium ion cells on lab tables in clean rooms that can completely charge in a few seconds, weigh a fraction of what previous cells weighed and perform all sorts of amazing feats.

However, these are all experimental batteries and have not yet been commercialized. The batteries we can buy and use today were developed years ago, sometimes decades ago, and have undergone extensive development and commercialization processes to reach our workbenches.

In a few years the properties described in the previous sections of this chapter will likely begin slowly changing and improving as we enter a new era of lithium batteries. That might be in the next 5 years, or it might be in the next 20 years. For now though, the above descriptions cover the batteries that are generally available for us to use today.

Chapter 4: Sourcing lithium battery cells

There are a number of different ways to find the lithium cells that you'll need for a DIY battery project. Lithium battery cells aren't cheap, and thus the main factors that will likely affect your decision will be cost and availability. While buying new cells is always the best (and safest) method, there are a number of options available. We'll cover many different options in this chapter.

Buying new cells

The first method, and the one I generally recommend, is to buy new cells from established vendors. Lithium battery cells are not cheap, but by buying new cells, you'll be sure that you're getting high quality, safe cells that will last as long as you expect them too. That might be a few hundred cycles or a few thousand, depending on the type of cells, but at least you aren't likely to have any surprises.

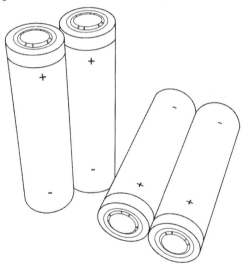

Depending where you live, you might not have a good local source for buying lithium batteries. Don't worry though - there are many sources to buy lithium batteries online. For years, lithium batteries were almost exclusively made in Asia, meaning that the best method to buy cells was to pay a stranger online and wait a month for them to arrive in the mail. Now there are an increasing number of resellers in countries around the world, making it easier to deal with a vendor that is at least located in the same country, even if they aren't local to your area.

I still mostly buy my battery cells straight from China, where you can usually get the best price. If you're buying small cells, like RC lipo packs or 18650 batteries, then buying online from halfway around the world can be a good option. Such lithium cells are usually small and can be shipped easily and cheaply. However, the constantly changing shipping regulations for lithium batteries might not make this true forever.

I generally buy boxes of 18650 cells by the hundred, which usually results in a pretty significant price reduction of around 10-30% compared to buying single 18650 cells.

A couple of the most common sources I use for buying smaller lithium battery cells are Alibaba and their retail division AliExpress. Alibaba is good for wholesale purchases where you'll be buying 100+ cells, though some vendors on Alibaba don't want to sell quantities of less than 1,000 cells. On AliExpress, you can buy cells anywhere from single units up to many thousands.

Always look for a vendor that has been in business for at least a few years and has lots of feedback ratings. It's true that good feedback can be bought or faked, but generally these vendors are selling in high enough volume that any bad feedback will still make it through. Vendors with hundreds of transactions and generally positive feedback are usually good sources.

There is a very large industry of fake lithium cells, which are usually off-brand cells that have been rebadged to look like brand name cells. When buying directly from Asia, it can be harder to determine if the vendor is selling genuine or counterfeit cells. Only buying from reputable vendors with years of good feedback helps stack the odds in your favor, but it's still hard to be 100% sure. Online forums for people building similar projects as you (such as electric bicycle or DIY powerwall forums) often have large communities that can help you find the most trustworthy dealers used by members of that industry.

If you're really worried about getting bad or fake cells, order just a few cells at first to test the vendor. That way you don't commit to a large order from a new vendor only to later find out that the cells aren't good. You could also consider just going with a local vendor that guarantees the authenticity of the cells they provide. It will be more expensive, but you can be sure that you're getting the right cells, and you'll presumably have recourse if you don't get what you paid for.

If you're buying large cells, such as those used in many electric vehicle conversions, your sources are going to be more limited. Most commonly available lithium batteries come in small form factors. Big cells exist, but there are fewer manufacturers and vendors. You'll likely do better by looking for a vendor in your country that can import large lithium cells directly for you. You can always use many small cells (Tesla electric vehicles have thousands of small cells), but for big projects like electric vehicle conversions or very large home energy storage batteries, large cells can save you a lot of work in the assembly phase.

Lastly, if you are going to spend the money to buy new cells, you should only buy name brand cells made by recognizable companies like Panasonic, Samsung, Sony, LG, etc. There are a number of off brands out there that sell cheaper lithium cells, but they don't have the same quality as the big players.

In the thousands of Panasonic and Samsung cells that I've purchased over the years, I've never once gotten a bad cell. The name brand companies all have excellent quality control. They test and remove any bad cells at the factory before shipment. Other companies have lower standards, and it's not uncommon to get a bad cell or two in a shipment of a few hundred. While this might seem negligible, it can actually have a devastating effect if you're building a large battery. A single weak cell can end up dragging down an entire parallel group connected to it, ruining many more cells.

There are also several brands of cells that simply buy rejected cells from the big companies and repackage them with ridiculously overrated specifications. Ultrafire is the poster child for this practice, but other brands such as Trustfire, Surefire and others do the same thing. If you see cells that are cheaper than name brand yet claim to have higher capacities, you can be sure they aren't legitimate cells. If it sounds too good to be true, it probably is.

Using salvaged cells

Another common source for lithium cells is to find salvaged cells. Lithium cells used in consumer electronics like laptops and power tools are rated for many hundreds or even thousands of cycles. This can be great when the products are designed to last, but so many products end up dying prematurely for various reasons. The lithium battery cells in these products are often then discarded along with the electronics long before the cells are actually dead. A laptop may die after a few years, but the batteries might have only gone through 20% of their rated charge cycles.

Many scrupulous DIYers have found that they can collect these discarded OEM batteries, open them up and salvage the lithium cells inside. This can be fairly labor intensive depending on how the cells are held in their OEM packaging and how many cells are included in each battery. However, the price simply can't be beat. Salvaged batteries are likely the cheapest way to get a large supply of battery cells.

Many of these discarded battery packs can be acquired for free from businesses that accumulate huge quantities, such as computer repair stores, recycling centers, hospitals and clinics that use battery powered devices, etc. In many places it is technically illegal to throw away these lithium battery packs in normal municipal garbage, and so a lot of these businesses actually pay companies to come and take these battery packs away for recycling. By offering to take the batteries off of their hands, you are saving them the expense and headache of dealing with what they consider to be electronic waste.

It's like they always say, "one man's electronic waste is another man's treasure". Or something like that.

As this practice has become more common, some places have begun charging for their discarded battery packs, but the prices are still much cheaper than you'd spend on new cells. A few dollars per pound is a common figure for these bulk battery packs.

As I mentioned above, I always recommend buying new cells when possible. This is because there are a few main downsides to using salvaged cells.

1) You never know what quality you'll get from salvaged batteries. Are they name brand? Are they counterfeit or inferior quality cells? Did the previous owner abuse them?

2) You don't know how much capacity each salvaged battery cell has. When you use salvaged batteries, you need to go through and check every single cell for capacity to determine which cells hold a charge and have reasonable battery capacity. Old cells lose their capacity, so it's not uncommon to throw out over half of the salvaged cells from a single haul. The process of checking every cell for capacity can take anywhere from a single afternoon to many months, depending on how many cell testers you want to invest in.

3) You don't know how much life is left in each salvaged battery cell. Was the cell only used for 20% of its total rated cycles, or was it used for 90% of them? If you know the original rated capacity of the cell, you might be able to approximately gauge this by how close it still is to its rated capacity. However, many salvaged cells aren't marked with their original capacity or even any identifying information. That means you're left guessing. Furthermore, if you build a pack out of cells with different remaining life expectancies, the first cells to begin dying will start to drag down the rest of the cells with them. The good cells will have to pick up the slack, which means that the good cells will get overworked and die quickly as well.

Depending on your goals and requirements, these disadvantages might be more or less of an issue. If you want a powerful battery for an electric vehicle or electric bicycle, salvaged cells are not the best way to go. There are simply too many unknowns. If you're building a backup battery for your house or to store energy from your solar panels and use it a reasonable rate, salvaged cells can be a great option. In fact, a huge portion of the DIY powerwall community uses salvaged cells exclusively in their battery projects.

The main thing to remember when using salvaged cells is that you should always use them conservatively - don't try to draw too much power out of them. It is better to build a larger capacity battery than you think you'll need in order to draw less current from each salvaged cell. We'll talk more about cell ratings and how much power cells can handle in Chapter 5.

Buying used electric vehicle battery modules

This option is something of a compromise between buying new, name brand cells and finding free or cheap salvaged cells. Used electric vehicle battery packs are becoming increasingly available on the second hand market through sites like eBay. With more electric vehicles on the road, more of these battery modules are available due to crashes, replacements and upgrades.

Buying used electric vehicle battery modules still means that you don't know exactly how much life is left in the batteries, but you at least know they are all at the same level of charge cycles (life expectancy) and that they are all name brand, genuine cells, assuming they came from a car that is known for having good cells. Most electric vehicles are made with good cells though - there's simply too much at stake to skimp on cell quality.

Some electric vehicles, like those made by Tesla, have batteries built from thousands of smaller cells. Other electric vehicles can have batteries made from just a few larger cells. Depending on what your requirements are, you'll have to decide which type of battery is better suited for your project.

Chapter 5: Cell ratings

Now that we've talked about different types of cells and where to get them, we need to learn about different cell specifications and what they mean. Just like crayons in the Crayola Big Box, there are a seemingly infinite number of lithium cells out there. Many of them look similar, but their specifications and ratings are what set them apart.

There's actually a very long list of different specifications and ratings for cells. Here we will look at the most important specifications that affect which cells you'll use for a given project.

Capacity

The capacity of a cell is probably the most critical factor, as it determines how much energy is available in the cell. Capacity of lithium battery cells is measured in amp hours (Ah) or sometimes milliamp hours (mAh) where 1 Ah = 1,000 mAh. Lithium battery cells can have anywhere from a few mAh to well over 100 Ah.

Occasionally the unit watt hours (Wh) will be listed on a cell instead of the amp hours. Watt hours are another unit of energy, but also consider voltage. To determine the amp hours in this case, simply divide the watt hours by the nominal voltage of the cell. For example, to calculate the capacity in amp hours of a lithium ion cell with a nominal voltage of 3.7 V and a capacity rating in watt hours of 11.1 Wh, simply divide 11.1 Wh by 3.7 V to get 3.0 Ah (or 3,000 mAh).

Amp hours of cell = watt hours ÷ nominal voltage of the cell

Or in our example above,

Amp hours of cell = 11.1 watt hours ÷ 3.7 volts = 3.0 amp hours

Capacity ratings only tell you how much energy the cell can store and provide. They don't give you any information about the power of the cell or its longevity. In fact, the highest capacity batteries usually have only moderate power levels. There is often a tradeoff between power rating and capacity. Therefore, the only thing you can use the capacity rating for is to determine how much energy is in a cell.

One last note about capacity - don't expect your cells to achieve their entire rated capacity. Manufacturers usually test the capacity of their cells using two "tricks" to eek out as much capacity as possible. They test at an incredibly low discharge rate, usually about 0.2 C (we'll talk about C rates later in this chapter). They also discharge the cells all the way down to their absolute minimum rated voltage, often 2.5 V for most li-ion cells. Discharging that low is possible, but it will decrease the lifetime of the cell if it is done too often. Most commercial products using li-ion cells discharge down to around 3.0 V, if not higher, in order to get longer life out of the cells.

So just because your cell might say it's rated 3.0 Ah, don't be surprised to see it perform closer to 2.9 Ah when you test it.

Maximum discharge rate

The maximum discharge rating tells you the maximum load, which is to say the maximum current, that can be drawn from the cell. There are actually two common discharge ratings, the "maximum continuous discharge current" and the "maximum peak discharge current". The maximum continuous discharge current is the better figure to use when making comparisons between cells. This is the maximum current that the cell can supply continuously without overheating or damaging itself. If the the maximum continuous current rating is 10 A, then the cell can provide 10 A of current continuously from its full charge state until its empty state.

The maximum peak discharge current is the amount of current that a cell can provide for a short burst. Every manufacturer rates this differently, which is why it is hard to use this number for comparison. Some manufacturers consider this to be a 2-3 second burst, while others consider a 10 second or longer period for the maximum peak discharge.

Regardless, you should never exceed the maximum peak discharge rating for more than a couple seconds, and if possible, you should try to design your battery to be sufficiently large that you never even reach the maximum peak discharge (we'll discuss battery design to minimize current load in Chapter 6).

Because manufacturers usually rate their cells at the extreme end of what they are capable of, it is never a good idea to push them to the limit of these ratings. Furthermore, battery cells that are operated near their rated limits tend to get very hot and operate inefficiently, robbing them of up to 10% of their designed capacity. So if you want to use the entire capacity of a battery cell, don't push it to its maximum discharge current limit.

C rate

The C rate of a battery cell is a measurement of the rate that the battery cell can be discharged or charged in relation to the cell's capacity. The C rate does not change based on the capacity of the battery cell; rather, it is an intrinsic property of the battery cell itself. That means that two cells that are identical in every way except for their rated capacities will also have identical C rates.

If that sounds confusing, then don't worry. We're going to work out some examples.

The C rate is calculated as a multiple of the capacity rating of the battery. A battery cell with a rated capacity of 2 Ah and a maximum continuous discharge current of 4 A has a C rate of 2. This would be known as a 2 C battery. In this example, we found the C rate by dividing the maximum discharge current rating by the capacity, which is calculated as:

C rate = 4 A ÷ 2 Ah = 2

This gives us a C rate of 2. If that same 2 Ah cell had a maximum discharge rating of 5 A instead of 4 A, it would be a 2.5 C battery. If it had a maximum discharge rating of 10 A, it would be a 5 C battery. If it had a maximum discharge rating of 10 A but it had a capacity of 5 Ah instead of 2 Ah, it would be back to a 2 C battery. Got it? If not, try working out those examples on paper using the same equation as above. We'll also have some more examples coming up soon.

The C rate is important because it is used to compare the relative power of cells, even when cells have different ratings. A big 10 Ah cell might be rated for 10 A maximum discharge while a smaller 2.5 Ah cell is only rated for 5 A maximum discharge. At first it might seem like the big cell is more powerful, as it can provide twice the current than the small cell can provide (10 A instead of 5 A). However, it is the smaller cell that is in fact more powerful relatively, as it has a higher C rating. The smaller cell has a C rating of 2 while the larger cell has a C rating of 1. If we combined four of the smaller cells together in parallel, we would make a 10 Ah battery that was rated at 20 A maximum discharge. Compare that to the original "bigger" 10 Ah battery cell and we can see that the bigger cell is in fact the weaker cell, as it is rated for only 10 A maximum discharge.

Occasionally lithium battery cells are marketed with just a C rating and not a maximum current rating. This can make it easier to compare the power level of battery cells of different capacities. As long as you know the capacity of the cell, you can use the C rate to quickly calculate the maximum current rating of the cell.

Maximum charge rate

The maximum charge rating is similar to the maximum discharge rating and is also fairly self explanatory - it's the maximum rate that you can charge the cell. Most cells will have a charge rating of not more than about 0.5 C. Charging a cell at a rate near its maximum charge rating will shorten the life expectancy of the cell. It is recommended not to charge most lithium cells at more than 0.5 C, and charging closer to 0.2 C is much better for the cell's health.

Remember how we calculated the C rating above for the discharge rate? It works the same for the charge rate. A 5 Ah cell charged at 0.5 C would be charged at 2.5 A. However, if you charged a 2.5 Ah cell at 2.5 A, that would be 1 C charging (and very fast charging as well, by lithium battery standards).

Maximum number of cycles

Depending on the type of lithium battery, the number of cycles could be anywhere from 200 to 3,000 or more. Cycle ratings can be difficult to compare from one cell to the next though, as manufacturer's don't always use a standard rating system.

The most common rating system is the number of cycles before a cell reaches 80% of its original rated capacity. The capacity of lithium cells slowly degrades over time with increasing charge cycles. At 80% of the original rated capacity, many manufacturers consider the cell to have reached the end of its useful life. Some manufacturers give a rating of charge cycles until 70%

capacity remaining. Some don't even specify, and just state a number of cycles until "end of life", leaving it unclear exactly what they consider the end of life to be.

Regardless, when a cell reaches 80% or even 70% of its original rated capacity, it isn't necessarily *dead*, it just isn't going to perform as well. Not only will it obviously not have as much capacity, but it will also have a larger "voltage sag", or drop in voltage under load. This will result in less power and an even shorter working life for the battery.

There are many other ratings and specifications for most cells, including everything from dimensions to temperature ranges. However, the ratings listed above: capacity, discharge and/or charge ratings and cycle life are generally the most important parameters for choosing battery cells for most projects and also help with comparison between multiple cells.

Chapter 6: Combining cells to make battery packs

Larger lithium battery packs are made by combining individual lithium battery cells. By combining multiple cells, different voltages and capacities can be achieved. The way these cells are combined determines the ultimate specifications for each resulting battery pack.

Increasing voltage using series connections

Individual lithium battery cells are usually 3.7 V nominal (for li-ion cells) or 3.2 V nominal (for LiFePO$_4$ cells). This voltage is acceptable for some low power devices, such as cellphones and other small electronics, but it can't provide enough power for anything more substantial. For bigger projects, including anything from an electric skateboard to an electric car, multiple lithium cells are wired in series to increase the voltage of the battery.

In a series connection, the positive terminal of one battery cell is connected to the negative terminal of the next battery. If you've ever slid more than one battery in a row into a flashlight tube, that is a series connection. The positive terminal of one cell always connects to the negative terminal of the next cell.

These series connections can combine just two cells or hundreds of cells. The number of cells wired in series depends on what voltage is required. To calculate the voltage of a set of battery cells connected in series, simply multiply the voltage of one cell by the number of cells in the series connection.

Total voltage of cells in series = nominal voltage of a single cell × # of cells in series

If we are using 3.7 V nominal li-ion cells and we connect two cells in series, we'll end up with a 7.4 V nominal battery calculated as follows:

Total voltage = 3.7 V × 2 cells = 7.4 V total

If we added one more cell into that series connection for a total of three cells, we'd have an 11.1 V battery. Ten cells in series would give us 37 V nominal. Fifteen cells in series would give us 55.5 V, and I think you get the idea from there.

$$\boxed{-\ \ 3.7\,V\ \ +}\ \boxed{-\ \ 3.7\,V\ \ +} = 7.4\,V$$

$$\boxed{-\ \ 3.7\,V\ \ +}\ \boxed{-\ \ 3.7\,V\ \ +}\ \boxed{-\ \ 3.7\,V\ \ +} = 11.1\,V$$

An important thing to remember is that the nominal voltage of a battery cell or a larger battery pack is just that - "nominal", which comes from the same root as the word "name". Basically, these cells are "named" 3.7 V cells, but in reality they span a much wider voltage range during use.

A single 3.7 V nominal lithium-ion cell can be charged up to 4.2 V and discharged as low as 2.5 V, which is a very big range. Image that we connect 10 of these cells in series to create a 37 V nominal battery. That battery's voltage will actually range from a fully charged voltage of 42 V down to a minimum of 25 V if discharged to 0% state of charge.

If we have a device that requires at least 30 V to operate, then we would stop discharging at 3.0 V per cell in this 10 cell battery, even though the battery could have kept discharging down to 25 V. That equates to not using about 5% of the pack's total capacity. You might not think 5% is a big deal, but what if that device required 35 V? We'd stop discharging at 3.5 V per cell in this 10 cell battery, which would equate to leaving about 40% of the battery's capacity unused. This is why it is important to consider the entire range of the voltage of a battery when calculating how many cells in series are required for your project.

Many electronics such as inverters, electric motors and other DC devices are designed for voltages in 12 V increments, such as a 12 V headlamp or a 48 V electric bicycle. This is a holdover from the many years when lead acid batteries were used to power these types of devices. Lead acid batteries use cells that have a nominal voltage of 2 V, and six are usually connected together in series to create 12 V lead acid batteries. Those 12 V batteries are then easily connected in series to create any other size battery with 12 V increments.

The problem this old system has created for us is that most lithium batteries don't conform well to this arbitrary 12 V increment.

Most electronics (but not all!) are capable of handling a small range of voltages above and below their rated voltage. For example, a 12 V LED headlamp can likely function with a voltage of between 9 V - 15 V, though more sensitive electronics will have smaller permissible voltage ranges.

This voltage range allows us to use a lithium battery voltage that is close to the 12 V increments that many electronics are rated for, even if it isn't exact. For example, electric bicycles are usually designed for either 24 V, 36 V or 48 V batteries. Again, this is because most of the ebike components were originally designed for lead acid batteries and the nomenclature in the industry just stuck.

The most commonly accepted lithium-ion battery for 24 V ebike systems is 7 cells in series, which creates a 25.9 V nominal battery that actually ranges from approximately 21 V - 29 V during use. For 36 V lithium batteries, nearly all ebike manufacturers use 10 cells in series to create a 37 V nominal battery that ranges from approximately 30 V - 42 V during use. When it comes to 48 V batteries though, the industry is fairly split. Batteries with 13 cells in series used to be the most popular configuration for a 48 V battery. This resulted in a nominal rating of 48.1 V and a voltage range under use of approximately 39 V - 54 V. However, with voltage sag, the battery would actually spend the majority of its time below 48V, which results in less power.

For this reason, many 48 V batteries for ebikes are now made with 14 cells in series which gives a nominal rating of 51.8 V and has a higher voltage range of approximately 42 V - 58.8 V. These batteries are often referred to as 52 V batteries instead of 48 V batteries to signify that they are indeed of a higher voltage than "standard" 48 V lithium batteries.

Other industries don't always have this 12 V increment issue and can essentially use any voltage that they design their devices for. Battery powered tools are a great example. Many power drills use 11.1 V nominal batteries which consist of three lithium-ion cells in series, though many tool manufacturers still call these 12 V batteries. This isn't really fair, as they spend very little time above 12 V. However, because they charge up to 12.6 V, the 12 V badge isn't technically untrue. Rather, it's just misleading. The next step up in power tools is usually an 18 V nominal battery, which consists of five lithium-ion cells in series.

One thing to note is that all of the above examples I gave used lithium-ion cells, as these are the most commonly used cells in these applications. However, LiFePO4 cells actually lend themselves more easily to 12 V increments. With 3.2 V nominal cells, combining four LiFePO4 cells will create a 12.8 V nominal battery, which is pretty darn close to 12 V. LiFePO4 cells are fairly popular for DIY electric vehicle conversions, largely due to a combination of high cycle life, excellent safety and only moderate space restrictions (who needs that trunk space anyways?). As many electric vehicle components were originally designed for lead acid batteries, they are also often rated in 12 V increments, which makes using LiFePO4 cells a bit easier when you're aiming for a specific voltage in a 12 V increment.

Increasing capacity using parallel connections

Combining cells in series increases the voltage, yet it has no effect at all on the capacity of the battery. Combining 10 li-ion cells that are each rated for 3.5 Ah in series will result in a 37V 3.5Ah battery. That's a decently high voltage, but very low capacity for most applications. In order to increase the capacity of the battery, we must combine cells in parallel.

Parallel connections can be thought of as the opposite of a series connection. Instead of connecting the positive terminal of one cell to the negative terminal of the next cell as in series connections, parallel connections are made by connecting the same terminals together. To connect two cells in parallel, you simply connect the positive terminal of the first cell to the positive terminal of the second cell, and then connect the negative terminal of the first cell to the negative terminal of the second cell. This essentially creates one larger cell, because the two cells are now sharing the same terminals and function as one battery cell.

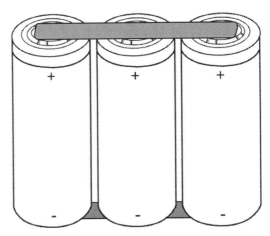

3 cells connected in parallel

One important safety note: before you connect any battery cells or battery packs in parallel, you MUST ensure that they have nearly identical voltages. If the voltages are off by a large amount, it means that one cell is at a higher state of charge than the other. When you connect cells with mismatched charge levels in parallel, the higher charged cell will try to charge the lower charged cell. If the difference in charge is large, the higher charged cell will try to dump a large amount of energy at once into the lower charged cell. This high current flow will damage both cells and can result in the cells overheating or catching on fire. Always check that cells are very close or identical in voltage before connecting them in parallel.

Ok, now let's look at some simple math. Calculating the total capacity in Ah of battery cells connected in parallel is easy: just multiply the number of cells connected in parallel by the capacity of the individual cells.

Capacity of parallel cells = the # of cells in parallel × capacity of a single cell in Ah

For example, let's say that we have two li-ion cells, each with a nominal voltage of 3.7 V and rated for 3.5 Ah. If we connect them in parallel by joining their positive terminals together and then their negative terminals together, we will have created a 3.7V 7.0Ah battery pack, as shown in the diagram. If we add one more cell in parallel with the first two, that will create a 3.7V 10.5Ah battery pack. If we connect 10 of these cells together in parallel, that will create a 3.7V 35Ah battery pack. Easy, right? Great!

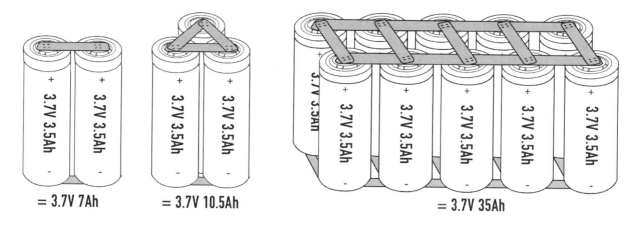

= 3.7V 7Ah   = 3.7V 10.5Ah   = 3.7V 35Ah

Combining series and parallel connections

We've now seen that series connections increase the voltage of a battery pack but don't affect the capacity, while parallel connections increase the capacity of the battery pack but don't affect the voltage. So how do we increase both the voltage and the capacity simultaneously? We simply combine both series and parallel connections together.

Let's look at an example. We'll start with parallel connections. Let's take those same 3.7V 3.5Ah li-ion cells from the previous example and wire three of them in parallel. This will effectively turn those three cells into one larger cell that maintains its 3.7 V nominal voltage but now has a combined capacity of 10.5 Ah, which can be calculated as:

Total capacity = 3 cells in parallel × 3.5 Ah per cell = 3 × 3.5 = 10.5 Ah

Now let's do it again. We'll take an additional three cells and combine them in parallel, just like the first three. Now we've got two small battery packs of three cells each. Both packs are 3.7 V nominal and have capacities of 10.5 Ah.

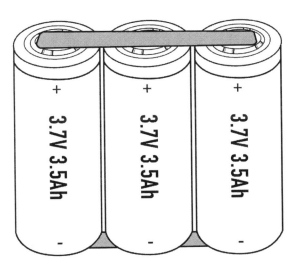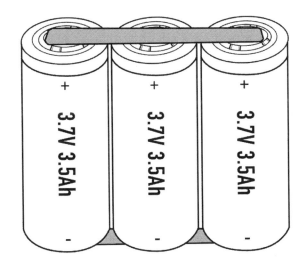

Alright, are you ready to go nuts? Now let's combine those two packs in series. That means we will electrically connect the positive set of terminals on the first three-cell group to the negative set of terminals on the second cell group. This connection can be made with wires, or with metal tabs, or even simply by pressing the terminals against each other. By joining these two parallel groups in this way, we've created a series connection between the two three-cell parallel groups.

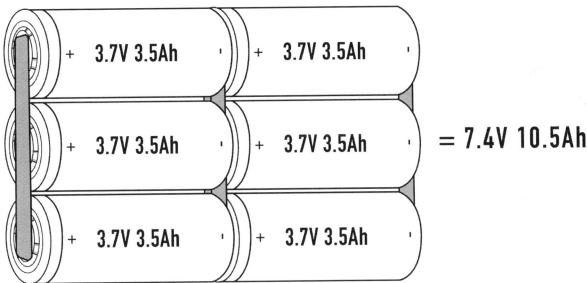

To calculate the new voltage of the combined battery packs we just multiply the voltage of each cell by the number of cell groups in series, which in this case is two (the two parallel groups each count as one big cell, not three individual cells, because they are connected in parallel and function as one cell). Our total voltage is then given by:

Total voltage = 3.7 V × 2 cell groups in series = 3.7 × 2 = 7.4 V.

Remember that the series connection that we just performed only affects the voltage, not the capacity. That means that our 10.5 Ah capacity remains the same. Therefore, by performing the parallel connections followed by the series connection, we have created a six-cell battery pack that is rated at 7.4 V nominal voltage and 10.5 Ah nominal capacity.

Now let's learn some terminology and abbreviations. Referring to the above battery as "three cells in parallel and two cells in series" is a mouthful, so we'll use the industry abbreviations of 's' for series and 'p' for parallel. The battery we created in the example above would be referred to as a 2s3p battery, as it has two series groups of three parallel cells each.

A 3s3p battery could be made by following the exact same instructions as the above example, except that we'd make three parallel groups in the beginning instead of two and then connect that third group in series as well. That 3s3p battery would be an 11.1V 10.5Ah battery. If we used the same cells to create a 10s4p battery, it would be rated at 37V 14Ah, which we can calculate as:

Total voltage = 10 cells in series × 3.7 V per cell = 37 V

and

Total capacity = 4 cells in parallel × 3.5 Ah per cell = 14 Ah

This series and parallel connection scheme is the basis of battery construction, so it is important that you completely understand it. Screwing up the series and parallel connections will not only result in getting your battery specifications wrong, but it can also very likely lead to a short circuit in the battery pack. Creating a short circuit in a lithium battery is a dangerous situation that we'll talk more about in Chapter 7. Suffice it to say, you want to get this stuff correct every time. To double check that you're good to go on series and parallel connections, let's do a little pop quiz.

Determine the nominal voltage and capacity of the following configurations:

A. 13s4p battery pack using 3.7V 3.0Ah li-ion cells
B. 8s2p battery pack using 3.2V 5.0Ah LiFePO$_4$ cells

Determine the pack configuration (the number of series and parallel cells) for the following specifications:

C. 37V 11.6Ah battery pack using 3.7V 2.9Ah li-ion cells
D. 51.8V 100Ah battery pack using 3.7V 10Ah li-ion cells

Determine the capacity of the cells used in a battery pack of the following specifications:

E. 25.9V 17.1Ah battery pack of configuration 7s6p built using 3.7V li-ion cells
F. 74V 20Ah battery pack of configuration 20s8p built using 3.7V li-ion cells

Answers: A. 48.1V 12Ah; B. 25.6V 10Ah; C. 10s4p; D. 14s10p; E. 2.85Ah cells; F. 2.5Ah cells

Chapter 7: Safety

Proper care must be taken when handling and working with lithium batteries and cells. If used properly, they can be quite safe. But if misused, they can present a serious fire hazard. Let's take a look at how things can go wrong with lithium batteries in order to learn what we should do right.

The danger of short circuits

Perhaps the best way to cause a lithium battery to catch fire is to short circuit it. A short circuit occurs when the two terminals of a battery pack or single cell are connected together. Essentially, a battery is put in series with itself, connecting its positive terminal to its own negative terminal. This creates a short loop of current that flows directly through battery with nothing to slow it down other than the battery's own low internal resistance.

A battery should never be short circuited. This is rarely done on purpose (outside of perhaps trying to light an emergency fire with a battery and some steel wool or a gum wrapper). Rather, most battery short circuits are the result of a careless mistake. These accidents can be incredibly costly. Best case scenario: the short circuit only occurs for a fraction of a second and results in only minor damage to the battery. Worst case scenario: the short circuit is maintained, causing the battery to overheat, catch fire and start a chain reaction with other cells or batteries around it, eventually burning down your home, car, airplane or anything else containing the battery.

So by now I think the message on short circuits is pretty clear: *don't do it*. You might be thinking, "that's easy, I won't do it." But it's not that simple. Like I said, most short circuits occur accidentally, usually due to carelessness or absentmindedness, and often during the assembly process.

Remember how we talked about series connections, where the positive terminal of one cell is connected to the negative terminal of the next cell? Well, performing series connections is a great opportunity to accidentally short circuit a battery. If your battery cells are lined up end to end, like in a flashlight tube, it's a bit harder to accidentally create a short circuit as you'd have to find something conductive that could span the length of the cells and reach around their ends to contact both terminals.

But imagine if we rearranged those cells so that instead of being in a straight line, they were next to each other with one cell flipped upside down. We could create a simple series connection by connecting the terminals on one end since the positive and negative terminals are right next to each other. But now look at the other end of the battery, it has a positive and negative terminal right next to each other, separated by just a few millimeters. Any metallic object large enough to reach both terminals could easily short circuit this battery.

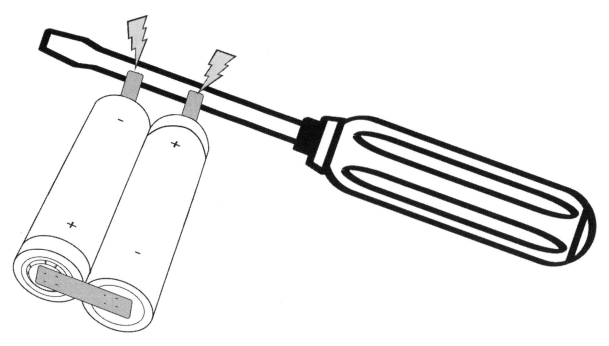

Imagine this battery sitting on your workbench. If you dropped a piece of exposed wire or solder onto those terminals, you'd create a short circuit. If you laid the battery down on a screwdriver, pliers or any other metal tool, you'd create a short circuit. You could short circuit that battery by touching your ring to it when you pick up that battery and hold it in your hand. I'm even speaking from experience on that last one: a gold wedding band makes an excellent conductor and leaves a very uncomfortable burn ring.

For that reason, you should always wear gloves when working with lithium batteries. Even if you don't have jewelry on your hands, sweaty palms can easily conduct current and you'll get quite a shock (literally and figuratively) the first time you experience it. Mechanics gloves work well, but I sometimes like to use latex or nitrile gloves as they give me better dexterity. Also, speaking of jewelry, remember to remove everything metallic. Watches, necklaces, earrings, nose rings, etc., basically anything metallic that hangs or dangles over exposed battery connections can be a hazard.

Now that's just a simple two-cell battery. Imagine a larger battery with dozens of cells and many more series connections. Every non-series connection between cell groups is another opportunity to accidentally short the cells in a battery like this. I don't want to make this sound too scary, because it shouldn't be, but it is important to keep in mind that the potential for short circuits is there during different stages of a build. From the moment you begin creating series connections until the moment the battery is sealed and the terminals are no longer exposed, there is the potential to create an accidental short circuit.

In order to protect against these kinds of accidental short circuits, remember to always work on a clean, clear surface. If your workbench is anything like mine, it's constantly cluttered and covered with bits and pieces of the last few (or more than a few) projects you've been working on. That seems to work fine for most projects, but it's an accident waiting to happen when you're working with lithium batteries. A single small screw or staple on your workbench can cause a short circuit with potentially disastrous results. Anytime you work with lithium batteries, completely clear off your work station.

I have a roll of cheap butcher paper and lay down a big piece on top of my work bench anytime I'm working with lithium batteries. Not only does it ensure that I'm starting with a surface free of any metallic hazards, but the white paper also makes it easy to see any new hazards I create, like drops of solder or snipped wire ends that could short my battery if I accidentally set it down on top of one of them.

But short circuits don't only happen from accidental contact with foreign objects bridging between neighboring cell terminals. Many short circuits occur from careless wire placement. When wiring your battery, you'll likely have multiple lengths of wire flopping around at different times on your work surface. These wires will be connected to different parts of your battery and have different voltage potentials, meaning that contact between the ends of these wires can cause a short. This happens more easily than you'd think, as many people are so focused on their battery that they don't pay attention to loose wires moving around on their workbench.

This is why you should never leave exposed ends on any wires that are connected to your battery. When possible, I like to attach connectors to my wires first before I ever attach them to my battery. This means that the ends of the wires are always covered from the second they are attached to my battery. If this isn't possible, I at least try to put a piece of electrical tape or heat shrink over the ends of the wires. If you're doing repair or maintenance work and have to snip wires, always cover the ends immediately with tape or heat shrink.

Also, beware of the springiness of many wires. If you aren't holding onto a wire when you snip it, it can often spring back after it has been cut. All it takes is a little bad luck for it to land anywhere on your exposed battery terminals and create a short circuit. So always hold onto a wire that you are actively cutting. I like to hold with multiple fingers and cut in between my fingers. That way I can hold onto both ends of the wire once it has been cut. When this isn't possible though, always make sure you hold onto the end that remains connected to your battery, as it has a higher chance of causing a short.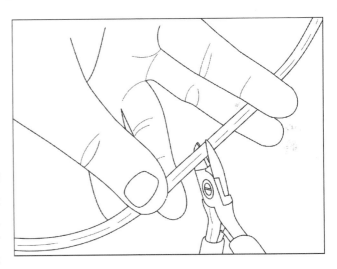

Another excellent way to short circuit your battery is by cutting multiple wires at the same time. When disassembling a battery or doing maintenance, you'll often need to use a wire cutter to snip off sections of wire or remove connectors. For the lazy among us, it can be tempting to snip multiple wires at once, but doing so can easily create a short circuit when two wires with different potentials both touch the metallic jaws of your wire snips at the same time.

Again, this may sound obvious to some, but it is easily overlooked, especially when working quickly. I'm not so proud that I can't admit that I have a pair of wire snips with a sizeable chunk of the jaws missing. They were vaporized from stupidly snipping off a charging connector at once instead of one wire at a time. In my defense, I was young and naive, but that excuse doesn't bring that tool back (or the few charge cycles it probably took off of my battery's life expectancy).

So always remember to cut wires individually. It can be annoying when you've got a dozen or more wires to snip, but it only takes a few more seconds and can prevent you from damaging your battery, your tools, or both.

Yet another great way to short circuit your battery is when you're adding connectors to your battery. If you don't add your connectors to the wire before it is attached to your battery then you'll be working with potentially "hot" wires, meaning the wires have the potential for current to flow if the circuit is completed. That is why it is best to add your connectors before you attach the wires to the battery.

However, sometimes adding connectors first just isn't possible. If you need to add connectors after the wires are already attached to the battery, add each connector one at a time and make sure to install whatever type of insulating shell comes with your connector, if necessary. If you have connectors such as bullet connectors, Anderson Powerpole connectors or others that require soldering or crimping the bare metallic connector first followed by the insulator, then it is even more critical to do these connections one at a time. Otherwise, you'll have multiple exposed connectors dancing out in space together, just asking for a short circuit.

Again, I'm unfortunately speaking from experience on this one. Many years ago when I was just getting started with batteries, I soldered two 6 mm gold plated bullet connectors onto the discharge wires of a 48V 15Ah battery before adding their insulators. I let one drop and it must have contacted the other. I say "must have" because I didn't see anything other than an intense ball of lightning. In fact, I never even saw the connectors again. They simply vaporized. They were gone. What had just a few seconds earlier been a 1 inch (2.5 cm) electrical connector had turned into just gas and dust before my temporarily blinded eyes.

So please, learn from my mistake. Don't let bare connectors dangle freely. Don't cut multiple wires at once. Don't handle an exposed battery with bare hands and while wearing a wedding ring. These all sound like stupid mistakes. And they are stupid mistakes. But they're stupid mistakes that are easy to make when we lose focus. All it takes is a few seconds of letting your mind drift to allow you to accidentally make a mistake.

When you're working with lithium batteries, and especially when performing connections between multiple battery packs and/or cells, there is no room for a lack of focus. Always think about what you're doing. In carpentry we say "measure twice, cut once". In battery building, the saying should be "think twice, act once". Whenever you're making a connection, cutting a wire or else performing any other type of action on a lithium battery, double check that you're making the right action, and then do it.

You'll notice that I devoted a lot of time to talking about short circuits. That is because short circuits are a convenient combination of both the easiest and most dangerous ways to go wrong with lithium batteries. But short circuits aren't the only ways to screw up and create a dangerous situation. Like a morbid version of Ron Popeil, I'm here to say "but wait, there's more!"

The effect of temperature on lithium cells

Heat is the enemy of lithium battery cells. Moderately high heat will cause your batteries to operate less efficiently, meaning your cells will die sooner and not deliver their full rated capacities. Moderately high heat should be avoided when possible, but it isn't necessarily dangerous. Lithium batteries can discharge under temperatures as high as 60°C, though it's better to keep that at a lower temperature, if possible, to increase their cycle life. Lithium batteries should never be charged at temperatures higher than 40°C, but we'll talk about charging specifics more in Chapter 13.

The actual danger occurs when lithium cells are brought to much higher temperatures. Lithium battery cells should never exceed 130°C. At temperatures slightly higher than 130°C, some lithium battery cells risk undergoing thermal runaway, which is when the electrolyte in the cells oxidizes at a rate that creates so much heat that it increases its own rate of oxidization, essentially fueling its own fire. As the cell grows hotter under thermal runaway, any nearby cells can also heat up to the point of thermal runaway, causing a chain reaction limited only by the number of cells nearby. Once a cell reaches thermal runaway, nothing can be done to stop it. The cell will combust until it has consumed itself.

The temperature at which thermal runaway begins will vary from cell to cell. Lithium cobalt cells can enter thermal runaway at temperatures as low as 150°C, while NMC cells generally reach thermal runaway at closer to 180°C. Both chemistries can reach temperatures of over 500°C at the peak of thermal runaway. The onset of thermal runaway varies greatly for $LiFePO_4$ cells, but some will begin thermal runaway at around 200°C, and most won't peak at higher than approximately 230°C during thermal runaway.

The physical effect on the cells largely depends on the type of cell. Cylindrical cells like 18650 cells have a vent mechanism on the positive terminal of the cell that allows gas to escape when the cell overheats and nears thermal runaway. Some prismatic cells have built in venting mechanisms. Other prismatic cells and all pouch cells don't include vents and will have no way to release the buildup of pressure in the cell. If the pressure exceeds the strength of the cell wall, these cells can rupture, often violently. Up to half of the gas venting from lithium battery cells is made up of highly flammable gases including hydrogen, methane and ethylene gas.

It is rare for cells to heat up to the the point of thermal runaway during normal use or storage. Your attic likely doesn't reach 130°C, though it's still better to store lithium batteries in a cool place at room temperature. Storing in a cooler location helps to increase the useful life of lithium cells.

The bigger risk of thermal runaway is when using lithium batteries under large loads that result in a high current draw on the individual cells. If the current is larger than the cell can handle, it will begin to heat up. If this goes on for too long, the cell can reach thermal runaway. For this reason, it is always important to design your battery to operate within the design specifications of the cells you are using. If you have a 20 A continuous load and are building a battery with four cells in parallel, those cells better be rated for at least 5 A continuous. Preferably, the cells would be rated for even more than 5 A continuous to ensure that you aren't pushing the cells to the limit.

On the other end of the spectrum, extreme cold temperatures aren't good for lithium cells either. Most cells can discharge in temperatures down to about -20°C, but they should never be charged at temperatures below 0°C. But again, we'll talk more about charging in Chapter 13.

Handling and storage

Depending on the type of battery cell, lithium batteries can be either surprisingly robust or incredibly fragile. Cylindrical cells are usually fairly strong due to their rigid metal case and can actually stand up to quite a bit of abuse. You shouldn't toss them around willy-nilly, but a few cylindrical cells bouncing around in a box isn't going to destroy them. When possible, you should try to keep them in some kind of plastic (not metal!) enclosure when you aren't using them, but they basically come in their own protective cases for all intents and purposes.

Also, with their positive and negative terminals on opposite sides of the cell, it's pretty hard to accidentally short cylindrical cells by having them loose in a box or a bag. Still, it's best to enclose them in some kind of case, just to be safe.

Prismatic cells can be fairly strong, depending on how they are made. Larger cells in the 20 Ah to 100 Ah range that are designed for electric vehicles or home/off-grid energy storage are usually encased in a fairly substantial enclosure. You should avoid dropping them from any decent height, but they generally aren't too fragile.

Pouch cells, on the other hand, need to be handled with care. They can't be tossed in a box like cylindrical cells can. Not only do they lack any protection from being stabbed, crushed, folded or torn, but they also generally have their positive and negative terminals on the same side of the cell, making it dangerously easy to short circuit them if they are stored in or near any metal objects. A loose paperclip could burn down a house if it shorts a couple pouch cells in a desk drawer.

Pouch cells are generally shipped and sold in formed plastic sleeves or cases. These plastic holders should be used to store the cells safely when they aren't being used. Never stack a big pile of pouch cells on top of each other unless you're actively connecting them and are both careful and aware of what you're doing. It's just too easy to accidentally short the wide, flat tabs of pouch cells on each other when they are left in a stack.

To summarize, lithium battery cells are inherently dangerous by design. They contain a large amount of energy in a small package and are designed to deliver that energy quickly. But by using proper precautions and safe operating principles, we can ensure that the cells are as safe as possible for our uses. Always pay attention to what you're doing and don't work with lithium battery cells unless you can devote your full focus to your work.

Chapter 8: Battery Management Systems

Battery Management Systems (BMSs), or as they are less commonly known, Protection Circuit Modules (PCMs) or Protection Circuit Boards (PCBs), are circuits that can be added to a lithium battery to protect the health of the individual cells in the battery. While optional, they are generally a good idea to include in most batteries.

Why would a battery need a BMS?

As we've already learned, battery packs are assembled by connecting multiple cells in series and parallel. And as we also learned, parallel connections between multiple cells make the cells act as if they were one big cell. If any one cell in a parallel group has a load applied to it, then all of the cells in that group experience the load equally. In this way, the voltage in all cells of a parallel group remains identical. As soon as one cell's voltage tries to drop, current flows in from its neighboring cells to maintain the voltage equilibrium. This is known as balancing. Cells that are balanced with respect to each other all have the same voltage (and thus charge state) as each other.

But think about those series connections. They don't work the same way. When a load is applied to multiple cells connected in series, all cells in the series chain will experience that load fairly equally. However, because the cells in series aren't connected at both terminals to each other (as that would make them parallel cells instead), they can't balance each other.

Now notice in the previous paragraph that I said those cells in series will experience a load *fairly* equally. I didn't say identically, and that is important. Small differences between the connections to the cells as well as the cells themselves will mean that each cell has a slightly different resistance. That small resistance means that a slightly different amount of current will flow through them as different amounts of current are burned off as heat in the form of resistance. The same current passes through the loop of series cells, but some cells burn off a little more or less of that current as heat. That results in slightly different states of charges for each cell or parallel cell group in a series chain.

Good quality cells with strong electrical connections will have nearly identical resistances and thus they will stay very closely balanced. However, at higher current levels, even good quality cells will eventually become unbalanced after dozens or hundreds of charge cycles. For low quality cells with large differences in internal resistance, or even for batteries made with good quality cells but with poor electrical connections, the cells can begin to lose their balance after only a few charge cycles.

The problem with becoming unbalanced is that unless you are using a specific charger known as a balancing charger to monitor every cell group and actively balance it during a charge cycle, then the cells will become charged to different levels. This is because simple, non-balancing chargers don't monitor the individual cell groups when they charge a battery. They just supply a voltage and a current, then wait for the battery to charge up to that voltage, at which point charging is completed.

With a non-balancing charger and an unbalanced battery, some cells will inevitably become overcharged while other cells don't get charged all the way. This results in multiple problems for the battery. First, the lower voltage cells have a lower state of charge because they weren't completely charged. Those cells will get drained even lower the next time the pack discharges, causing irreparable damage to those cells. Next, the higher voltage cells will have charged over 4.2 V (the proper full voltage for li-ion) or 3.65 V (the proper full voltage for LiFePO$_4$). Spending any amount of time over the maximum rated voltage will also cause irreparable harm to the cells.

Unbalancing of cells is also a runaway condition. As cells become unbalanced, they support a disproportionate amount of the load, which causes them to become further unbalanced, and the vicious cycle continues, increasing the imbalance until the battery destroys itself.

Alright, I know, you get it. Unbalanced cells are a bad thing. So what can we do to prevent cells from becoming unbalanced after many charge cycles?

One option is to use a balancing charger, like mentioned above. Balance chargers are commonly used for RC lipo batteries in the RC vehicle world where they are considered a necessity. Most balance chargers start by charging a lithium battery the simple way, by providing the appropriate total voltage and current to the entire pack across all cells evenly. This usually occurs by plugging directly into the discharge wires and reversing the current to use the discharge wires for charging. But balance chargers also connect to smaller wires (known as balance wires) that in turn connect to each individual cell in the battery (or each parallel group, if there are multiple cells in parallel). The charger monitors each cell (or parallel group) compared to the others and can drain or bleed off voltage from the higher voltage cells until all cells are balanced.

Most balance chargers are capable of performing both this balancing charging scheme as well as just the simple, "total voltage to the entire pack without balancing" scheme. The latter is known as "bulk" charging. In bulk charging, the cells do not get balanced and the issue of some cells becoming slightly overcharged while other cells become slightly undercharged still exists. For good quality battery cells though, the amount of unbalancing that occurs during normal discharge and charge cycles is quite small, meaning that the batteries can reasonably go a few or even dozens of charge cycles using bulk charging before they need to be balanced charged. The advantage of bulk charging is that it is faster - you don't have to wait for the cells to balance at the end of the charging operation. The disadvantage is of course that the cells will slowly become unbalanced. Depending on the specific battery, project requirements and user requirements, the correct ratio of bulk to balance charging can be selected.

Ok, you're probably wondering why I haven't talked about BMSs in a while. This is after all the BMS chapter. Well, you need the background theory to understand why BMSs are so important and useful. Now that you understand where we've been, let's look at where we going. Enter the mighty BMS!

The job of a BMS is to allow a simple bulk charger to be used to supply the correct total voltage to a battery while still making sure the cells don't become unbalanced. The BMS sits between the bulk charger and the battery cells and regulates the cells. It allows the total pack voltage from the charger to pass through the cells and charge them to the correct voltage. But during that charging, the BMS constantly monitors the cells and will start to drain off some energy from any cell or parallel group that starts to get charged too high.

The BMS basically does the job of a balance charger, but instead of residing in the charger, it usually sits inside the battery in the form of a separate circuit board. Larger BMS units, on the other hand, often have their own separate enclosures to help disperse the heat caused by the balancing process.

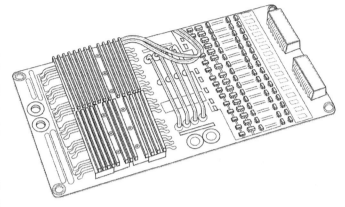

The type of balancing described above is known as 'top balancing' because it balances the cells at the top of the charge curves. This is the most common type of BMS balancing scheme. An alternative is known as 'bottom balancing' and it works just like it sounds. Instead of balancing cells when the cells are full, it balances them when they are approaching their empty state of charge. There's a lot of debate in the battery industry about which form of balancing is better. For the most part though, top balancing is the industry standard and what you'll find in most BMSs.

The fact that the BMS is always connected to the battery provides some significant benefits. BMS designers have taken advantage of this to develop BMSs with many additional features. Nearly all BMSs have protection circuitry for discharging as well. Because they are connected to every cell or cell group, they can monitor the voltage of every cell. Whenever the first cell hits the Low Voltage Cutoff (LVC), the BMS will cut the discharge circuit and stop the battery from being discharged any deeper. This protects the cells in the battery from over discharging and suffering irreparable damage.

Interestingly, some of the cheapest BMS's only have this Low Voltage Cutoff (LVC) feature, but not a balancing feature. This is something to watch out for if you're buying a budget BMS. Always double check with the vendor that it does in fact include the balancing circuitry.

Many BMS's also protect the pack from over-drain, which is where the BMS cuts power from the battery if the load on the battery exceeds a certain threshold. This protects the cells from working too hard and damaging themselves. It can also protect against short circuiting the battery, depending on the BMS.

Some BMS's have thermal protection built-in, where an included temperature probe monitors the instantaneous temperature and can stop a discharge or charge process if the battery becomes too hot. The temperature sensor is often on a length of wire allowing the user to place it against the cells of the battery where it will be more responsive to a sudden increase in cell temperature.

More expensive BMSs can include features such as Bluetooth connectivity and anti-theft or alarm features. Bluetooth connectivity allows the user to remotely monitor the status and health of the cells or entire battery pack in real time from their cell phone or other Bluetooth-enabled device. Anti-theft features can prevent the battery from providing current until a signal is received. Some BMSs include a display screen to give an accurate readout of battery specifications. This is a rare feature though, and most BMSs use a simple LED indicator to display whether or not charging of a cell has completed, if they have an indicator at all. There are also programmable BMSs that allow the end user to update the BMS's settings at his or her leisure, though these BMSs are considerably more expensive.

Ultimately, most applications only require a simple BMS that will protect the cells during charging and discharging. If you'd like something fancier though, there are many different BMSs out there to choose from.

Disadvantages of a BMS

In the previous section we talked about the advantages of using a BMS instead of a balance charger. However, there are also some downsides to using BMSs.

One of the main disadvantages is that a BMS can occasionally fail, and when it does, it will often cause the very problem it was designed to avoid. A BMS that fails can slowly drain the cells in a battery, often at a rate that makes it unnoticeable to the user. This can result in the cells becoming degraded or destroyed over time (or quite quickly, depending on the type of failure). Batteries that are mass produced with an emphasis on cost reduction are more likely to have these problems due to cheap BMS units with poor quality control standards.

While it is a rare problem to have, the ebike industry has seen a fair number of these cases. Manufacturers have sometimes tried to reduce the cost of the already expensive batteries by using cheaper BMSs or skimping on quality control. This has resulted in some batteries suffering a premature demise when their cheap or low quality BMS fails.

For this reason, BMSs have gotten something of a bad reputation in certain circles. Like anything, you get what you pay for, and if you want a good quality BMS, you should expect to pay a reasonable price. If a deal looks too good to be true, then it probably is. A cheap BMS can end up costing a lot of money in the long run if it destroys an expensive battery.

BMSs also limit the maximum amount of power that a battery can provide. Because the discharge path of a battery has to pass through a BMS (to allow the BMS to monitor and cut off discharge when necessary), the discharge current limit of a battery with a BMS is limited to the current limit of the components used in the BMS. Better quality BMSs can have very high discharge rates, but their prices rise accordingly.

This reason, along with the extra weight of a BMS, are the main two reasons why the RC vehicle industry has not adopted BMSs in their batteries. As dangerous as RC lipo batteries can be, a BMS would do wonders for helping monitor and protect such batteries. However, RC vehicles, and RC aircraft especially, all require a combination of very high power and low weight. A BMS both adds weight and reduces the amount of power that the battery can provide.

Interestingly, some people that use RC lipo batteries outside of the RC industry, often due to their wide availability and low price, will add a BMS unit to these batteries to increase their safety. When the high power and minimized weight requirements are relaxed a bit, a BMS becomes a welcome addition to such potentially dangerous batteries.

Connecting a BMS

Proper wiring between a BMS and a battery will be covered in detail in Chapter 11, so for now I'll just give you an overview so you understand how the process works.

First of all, it is important to match the right type of BMS for your battery cells. A BMS for a lithium battery will be designed for either li-ion cells or for LiFePO4 cells, but not both. It is extremely important that you use the right BMS for your cells. Mismatching the BMS and cell types could result in overcharging the battery. Overcharged lithium batteries are never a good thing. They have a tendency to go all 'fireworks show' on you. So make sure you choose the correct BMS!

Next, a BMS is generally the last thing added to a battery during the construction phase. I usually put mine on right before sealing up my battery. The reason for this is that the BMS is usually connected onto the inter-cell series or parallel connections. This avoids having to connect it directly to the cells themselves, which if you're using a soldering method, prevents unnecessary heat from reaching the cells. And as we already learned, excess heat is the mortal enemy of lithium batteries.

The BMS itself will have small wires, often referred to as "sense" or "balance" wires, and thicker wires (or pads to solder on thicker wires) which are used for actual charging and discharging of the battery. Those thick wires will get connected to the battery's main positive and negative terminals (or often just the negative terminal) while the thin balance wires will be connected between every set of parallel groups.

Some BMSs have an equal number of balance wires to the number of cells they are meant for (e.g. a 10s BMS with 10 thin balance wires), but it is also common to see BMSs with one more balance wire than the number of cells they are rated for (e.g. a 10s BMS with 11 thin balance wires). If you have an equal number of cells and wires, then the balance wires each get connected to the positive terminal of each parallel group. So for a 10s battery and 10s BMS with 10 wires, each wire would get connected to 1+, 2+, 3+, etc., all the way up to 10+. For a 10s BMS with 11 balance wires, that first wire will get connected to the negative terminal of the first parallel group, which is also the negative terminal of the entire battery. Then the rest would continue on like above, with the second wire connection to 1+, the third wire connecting to 2+, etc., all the way up through the 11th wire connecting to 10+.

But you don't have to worry about that too much now. That's all still theory for us. We'll circle back around in Chapter 11 and talk about BMS connections in much more detail when we discuss the actual assembling of batteries.

Chapter 9: Construction methods

Alright, so now we know all about lithium battery cells, how they work, what to do with them, and how to keep them safe. But now you are probably asking, "how do we physically combine them into a larger battery?" And if you weren't, well then now you're at least thinking about it. Well buckle up and hang on tight, because we've got a whirlwind of options and information here.

Physically holding cells together

The first thing we need to discuss is how to even hold cells together in a pack. Forget the electrical connections - we'll get to that in another couple pages. But how do we even keep the cells together? The answer largely depends on the type of cells you're using.

Cylindrical cells are both the most common types of cells for custom battery building and also some of the easiest to secure, so let's start there. All standard cylindrical lithium battery cells have commercials cell holders that have been produced to help people like you and me combine them into larger battery packs. For smaller cells like 18650s, these often come in the form of plastic blocks with circular holes in the middle. They snap together like Legos and make it easy to build two matching plastic racks that can hold each end of the cells. These blocks are usually black, but some larger cylindrical cells such as Headway cells like the 38120s have larger orange cell holders.

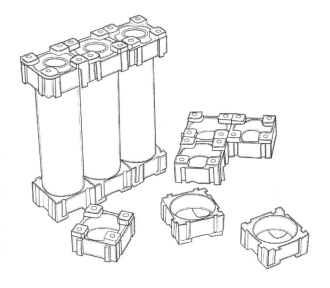

Regardless of the color or what they're made from, they're all pretty much the same. They generally come in either single cell blocks or larger rows or matrices that save you time when building larger batteries.

If you have access to a 3D printer and a CAD program (or can download other people's CAD files from the internet), you can even 3D print your own cell holders in any custom shape you want.

The advantages of these cell holders are mostly the convenience that they provide. They hold your cells nice and firm, making it easy to create your connections. They can also increase the safety of your pack-building process because they generally have small posts on the end that help keep the terminals of the batteries lifted above your work surface. This helps prevent accidental short circuits that could be caused by laying your partially completed battery down on a metal object.

Depending on the type of cell holders, they can also add some good rigidity to your battery. The cheap ones are, well, cheap - they don't really add too much strength because their tolerances are low and they snap together fairly loosely. The blocks that come in larger matrixes are usually stronger since they don't have individual joints between every cell.

If you are building a battery that will see a lot of movement and stress, such as for a skateboard or electric bicycle, strong cell holders can help strengthen your battery. Unless you're going to 3D print a custom shape to fit your application (or unless a rectangular shape works for you), then snap together cell holders will limit your battery design options.

Another benefit of cell holders like this is that they allow for airflow between the cells, which can help cool cells during use. However, this only helps if you leave your battery unsealed. If you heat shrink your battery or keep it in a case, then the cooling effect of these cells holders will be almost zero. The issue is that even though there is an air gap between the cells, if the battery is sealed in a box or in heat shrink, that air simply heats up and then stays between the cells. You're basically creating a low-power oven. This is not necessarily a bad thing if your battery stays within normal operating temperature ranges. Batteries naturally heat up during use, and they can withstand those normal temperature increases. Just don't try to leave gaps between your cells and then seal up the battery while patting yourself on the back thinking you've got great cooling.

For applications where waterproofing is not an issue, like for home energy storage batteries or other batteries that are kept indoors, the added cooling benefit of a battery built with air gaps between the cells can be a nice advantage. But you'll also want to make sure that your batteries are safe and protected from everything, not just water. Accidents often happen in unforeseeable ways. A pet climbing on your battery shelf can easily start a fire by knocking over unsealed batteries. Just keep that in mind.

Another option for cylindrical cells is to simply hot glue the cells together into whatever shape fits your application. Hot glue might seem low tech, but it's actually a very good option for many applications. Hot gluing cells together gives you the most efficient use of space, since you aren't putting plastic blocks or connectors between each cell. If you are trying to minimize size and weight, hot glue is probably the best option.

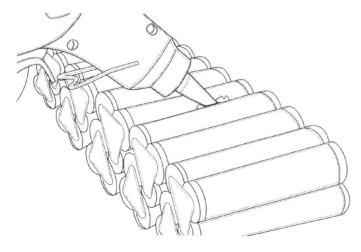

It also allows you to build any shape you want, as you can simply glue the cells together following any template you design to fit your application.

Hot gluing can be fairly rigid, but it isn't the strongest method to join cells together. It's limited by the strength of both the glue and the heat shrink on the cells. Because most cylindrical cells have a layer of heat shrink around the outside of their shell, the hot glue actually bonds to the heat shrink instead of the cell itself. That means that if you glue two cells together and then try to pull them apart, if the glue doesn't fail first then the heat shrink will eventually tear. Ideally you won't have this level of stress on your pack, but it is something to keep in mind.

Another disadvantage of hot gluing cells is that it creates a denser thermal mass. The cells in the center can't dissipate their heat to the air between cells because there is very little air available. As we discussed before, air gaps don't help much with cooling if the battery is sealed, but if the battery isn't being sealed and passive or active air cooling is available, then limiting the air gaps between cells by gluing them together will have a negative impact on the cooling rate of the battery.

For most batteries, the lack of air space around cells caused by hot gluing just won't be an issue. Batteries that are used for low to medium power applications simply won't create enough heat to cause a problem. Very large and high power battery packs will create more heat though, so this is something to consider.

Hot glue shouldn't loosen up during normal battery use conditions. Hot glue melts at around 120°C or 250°F, which is hotter than you want your battery to operate. However, if for some reason you do plan for your battery to be in an environment with abnormally high temperatures, remember that you might soften or lose your hot glue between your cells.

Some companies have experimented with wax blocks to hold cylindrical cells. AllCell Technologies, a company based in Chicago, uses a proprietary wax blend with cylindrical holes to hold 18650 cells. This wax is designed to undergo a phase change, shifting from solid to liquid form, as the battery cells begin to heat up. This helps pull the excess heat out of the cells. You can buy pre-built packs from AllCell, or you can find a number of homemade "phase change wax" recipes online to try and make your own version. A block of wax and a drill bit can result in a nice homemade battery holder!

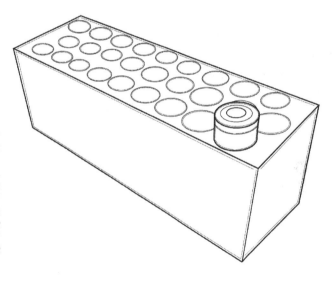

If you're using pouch cells instead of cylindrical cells, you'll need a much better solution for holding your cells. There aren't many commercially available cell holders available for pouch cells, mostly because there aren't any standard pouch cells sizes.

The amount of protection for your cells will depend on your requirements. RC lipo batteries are a great example of minimalist design. The pouch cells that comprise RC lipo packs are simply held together by a single piece of heat shrink. Occasionally there is a thin layer of plastic or fiberglass around the outside of the shells, but often a single layer of heat shrink is all that is used to combine and hold the cells.

The advantage of only using heat shrink is that you'll get the lightest and smallest battery possible. The disadvantage is that you've got very little protection. Any bumps or impacts are transferred directly to the actual skin of the lithium pouch cells. Anything sharp can easily puncture the cells and cause a devastating fire. This method also works better for smaller packs. As packs get larger, the increased weight and increased amount of contained energy require stronger and safer battery enclosures.

A hard plastic case is a better option for housing pouch cells. You can get fancy and build a custom box out of plastic or acrylic sheet, or you can use off the shelf products like OEM cases. Tool cases, lunch boxes and storage boxes that are made from hard plastic are great options. They are mass produced, which often keeps the cost down, and they are also already designed specifically for protection. A layer of foam on the inside of a hard case is a good idea to help cushion the cells and keep them from impacting the hard plastic directly.

Some people build wood enclosures for batteries. This is also an option, though I'd recommend using a soft foam layer between the wood and the cells, just like in a hard plastic case. The problem with wood is that not only is it more flammable, but if it is left unsealed then it can also absorb water if it ever gets wet. This could create a humid environment around your cells, which is less than ideal.

Metal cases are not usually a good idea for DIY batteries unless you have ensured that the battery and all connectors are 100% sealed and insulated, or you've painted or sealed the case to render it non-conductive. It's very easy to accidentally create a short circuit on a metal case if you aren't careful. Heat shrink wrapping the entire battery and using good, enclosed connectors can help make a metal case more viable.

Prismatic cells often come in their own hard plastic cases, which makes building a pack much easier process. Large prismatic cells in the 20 Ah - 100 Ah range are usually used in electric vehicle construction and can often be used as-is, no external cases required. They can be simply loaded into the trunk or engine compartment of a car, though a custom box to enclose the batteries and support them is a good idea. Such a box can also help take physical strain off of the electrical connections, busbars and wiring used to electrically connect the battery.

Prismatic cells are often held in wooden cases, usually because their large size makes it hard to find OEM plastic cases that can hold such large batteries. The fact that prismatic cells are sealed also makes it less of an issue if the wooden box were to absorb a little moisture. Smaller prismatic cells can be held in OEM plastic cases though, or just left exposed if they are built into their own rigid plastic cases.

Joining cells in series and parallel

There are a number of different ways to make electrical connections to join lithium cells together. The right method often depends on the type of cell and the load requirements on the battery.

Some lithium cells come with threaded terminals that make it easy to bolt the cells together. Many prismatic cells, especially the larger ones, have these threaded connectors. Sometimes they are threaded studs that extend from the cells and require a nut to be threaded onto the battery's stud terminals, and sometimes they are threaded holes that require bolts to be inserted into the battery case itself.

Headway cells are some of the only cylindrical cells that have threaded posts available for easy bolt-together pack assembly. Cylindrical cells usually have smooth, flat terminals that are designed to either be held by spring loaded battery contacts or spot welded.

Spring loaded battery contacts are the common kinds used for smaller batteries like AA's. They're the battery holders you often see in a remote control, and usually consist of a spring on one end and a button or rounded dome on the other. The spring ensures that the two ends of the battery are held in constant contact with the case terminals.

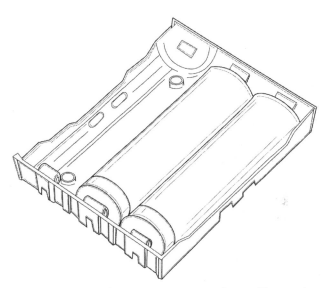

This method is probably the easiest way to join batteries into larger packs as it requires no tools and very little skill. Just line up the +'s and -'s and pop in the batteries! The problem with this method is that it only works for low power applications, which is why it is used in most consumer electronics. The spring contacts have relatively high resistance and so if you're drawing any significant amount of power out of the cell, you're going to wind up with a large voltage drop and a lot of wasted heat coming from the spring terminals. If you try to draw too much power, you can even overheat the spring contacts, which will begin to glow bright red. As you can imagine, this is bad. You want to avoid this.

One of the main advantages of lithium batteries is that they can provide quite high power. Taking advantage of this power means you can't use those convenient spring-contact holders.

Instead, the proper way to join lithium batteries that don't have screw terminals is by spot welding. Spot welding is a type of simple welding that works by passing a very high current through two pieces of clamped metal. An electrode on either side of the clamped metal (or in many battery welders, on the same side) press the metal together and applies a short pulse of high current. The current flows from one electrode to the first piece of metal, then through the second piece of metal and finally into the second electrode. The highest resistance is found between the two pieces of sandwiched metal. Because of the higher resistance between the metal, that joint quickly heats up first and causes the metal to instantly melt, joining the two pieces of metal into one at that point.

The key to spot welding is that the pulse of current is very short, often just a few milliseconds. That means that the only place that heats up substantially is the joint between the two metals being welded.

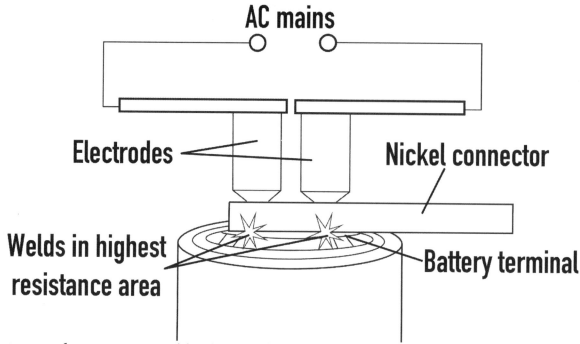

In fact, most battery spot welders use a double pulse, where two extremely short pulses of current fire in rapid succession. The first pulse often helps to soften the metal and burn off any impurities, while a second and usually stronger pulse performs the actual weld.

As we've talked about repeatedly, heat is the enemy of lithium battery cells. By using spot welding to join cells together, only a very small amount of heat is created for a very short time, and that small amount of heat quickly dissipates into the ambient air. This is crucial, because the more heat that is transferred into the lithium battery cell, the more damage will be caused to the cell.

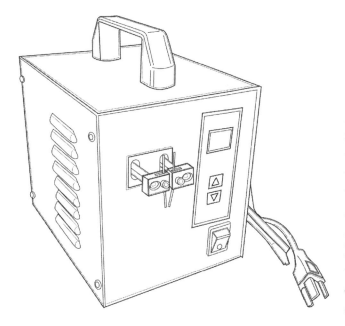

Spot welding is the most common method used for joining lithium battery cells that don't have screw terminals, and its the method used by nearly every professional battery builder.

Professional battery spot welders usually cost many thousands of dollars and are outside of the reach of DIY battery builders that only plan to build a few batteries. The professional spot welders often have CNC capabilities to automatically weld a preprogrammed pattern and usually use capacitive discharge. However, hobby-level battery spot welders that use transformers instead of capacitors are fairly common and can be found for a few hundred dollars on sites like AliExpress and eBay.

The cheapest hobby level spot welders cost around US $100, but usually leave a lot to be desired in the quality department. Even the $200-$300 spot welders are known for being a bit of a gamble sometimes. They usually either work great or arrive broken. And because they are often only available by purchasing them directly from Asian vendors online, it can be often be difficult to return a defective or broken unit.

There has been something of a renaissance in DIY spot welders based on microcomputers like the Arduino or Raspberry Pi. These spot welders often use a car battery as the current source and use microcontrollers to control the electrical pulse. While this option usually requires a handy operator to build their own spot welder, it often ultimately results in a better quality welder.

The most common material used for spot welding onto battery cells is nickel. Nickel is an excellent option for building electrical connections for battery packs because of its high conductivity and relatively low resistance. Copper is often used for electrical connections as it has even higher conductivity and lower resistance than nickel, but that same low resistance makes it more difficult to spot weld. The problem is that most spot welding electrodes are made from copper and if you remember how the spot welding process works, we need to create a higher resistance between the two pieces of metal being spot welded than between that same metal and the electrodes. If all of the parts are made from copper, it is difficult to ensure that the resistance between the two welded surfaces will be substantially higher than between the electrodes. Nickel, on the other hand, is quite easy to spot weld with copper electrodes yet still has low enough resistance to make it a good choice for electrical connections between battery cells.

Rolls of nickel strips are commonly available specifically for battery pack welding. Nickel strips are produced in many different widths and thicknesses. The maximum thickness of nickel that can be welded by most hobby-level spot welders is approximately 0.15 mm. For 18650 cells, you generally shouldn't use anything narrower than 7 mm, otherwise you'll be reducing the conductivity of your cell connections. I like to use 8 mm wide strips of 0.15 mm thickness, and will often use multiple strips welded on top of each other when a higher current battery is required.

Damian Rene, a custom battery builder in Madrid, calculated and released a handy table for determining the amount of current that can safely be carried by nickel strip of a certain size. The optimal range is where the nickel barely rises in heat above the ambient temperature. The acceptable range is where the nickel begins to heat up but does not excessively overheat. The poor range is where overheating occurs and currents above this level should be avoided.

Acceptable current levels for pure nickel strips			
Size	Optimal	Acceptable	Poor
0.1 mm x 5 mm	< 2.1 A	~ 3.0 A	> 4.2 A
0.1 mm x 7 mm	< 3.0 A	~ 4.5 A	> 6.0 A
0.15 mm x 7 mm	< 4.7 A	~ 7.0 A	> 9.4 A
0.2 mm x 7 mm	< 6.4 A	~ 9.6 A	> 12.8 A
0.3 mm x 7 mm	< 10 A	~ 15 A	> 20 A

Because I like to use 8 mm wide, 0.15 mm thick nickel strip, my general rule of thumb is that for this size, one strip can carry 5 A. If I need more than 5 A of current carrying capacity per strip, then I weld another strip on top of the first. I always weld one strip at a time to ensure that I get the best possible welds. As you can see from the table, the acceptable range for nickel strip of that size is somewhat higher than 5 A, but I use this as a rule of thumb to stay on the safe side.

It would also be possible to use thicker nickel strips instead of welding multiple layers, but most hobby grade spot welders can't handle nickel strips of more than around 0.15 mm in thickness. Welding two layers of nickel has the same effect as using one layer of nickel that is twice as thick.

Nickel strip is usually available in two varieties, pure nickel strip and nickel plated steel strips. Unless you have very low current requirements, you want pure nickel strip (which is often advertised as 99.9% pure nickel strip). Nickel plated steel strips are cheaper to produce but have much higher resistance and thus are only suitable for low power batteries.

Be careful when you purchase pure nickel strip to ensure that it is in fact pure nickel. When purchasing nickel strip from international vendors, I have often seen nickel plated steel passed off as pure nickel. The two look identical and are hard to distinguish, even when given a sample of each. Both have the same appearance and nearly identical densities (or close enough that you need laboratory grade equipment to measure them accurately enough to distinguish them by mass). Together, Damian Rene and I developed two tests to determine whether nickel strips are pure or steel plated.

The salt water test: Nickel does not corrode easily, unlike steel which loves to rust. Take a piece of nickel strip and scratch it up aggressively with strong sandpaper or a metal file. If you don't have either, use a coin or a nail or anything really - just make sure you really attack the surface. We want to scrape off some of the nickel layer, if it's there. Then mix up a cup of salt water (the exact amount doesn't really matter). Drop your scratched nickel strips in the water and wait a few days. If you don't see any rust forming, you've got pure nickel! If you see brown or orange rust, you've got nickel plated steel strips.

The sanding/grinding wheel test: I developed this test on accident while experimenting with the first test. I used a Dremel sanding wheel to scuff up the surface of a pure nickel strip and a nickel plated steel strip before doing the salt water test. The pure nickel strip was fairly uneventful, but when I used a sanding or grinding wheel on the nickel plated steel strip, I got brilliant sparks shooting from the strip as soon as the wheel ate through the thin nickel layer and contacted the steel. Makes sense - steel creates spark showers. Apparently though, nickel does not, and that makes this another great test. It's also superior to the salt water test because you get the result instantly. But if you're still in doubt, go ahead and toss the piece in salt water after you're done grinding on it, and that will verify the results of the grinding wheel test.

In addition to nickel strips, some people use nickel sheets for joining cells, especially when building larger battery packs. The advantage of nickel sheets is that you can cut them to any shape you want. Nickel strips require making straight line connections, but nickel sheets can be cut to abstract shapes and reach around to different parts of a battery. They also make it easier to use one layer of nickel as they are wider and can carry more current than individual nickel strips. We'll talk further about using multiple nickel strips to carry more current in Chapter 10.

Nickel is a great material for spot welding between cylindrical cells, but other types of cells don't require any extra material, depending on your battery configuration. Pouch cells have large battery terminals that can be joined in a number of different ways.

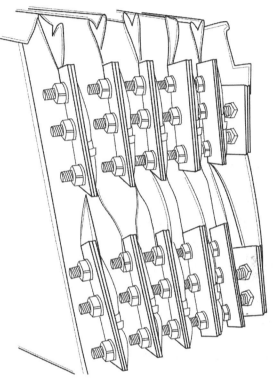

Some people use aluminum blocks to clamp onto the battery tabs, and then wires can be soldered or bolted directly to the aluminum blocks. Others have drilled holes in the tabs and bolted through the tabs to connect wires directly to the tabs themselves. Be careful with this method, as you want to be sure that you're only drilling through the tab. Clamping the tab to a jig with pre-drilled holes can be a good method to ensure that your drill bit doesn't walk and accidentally puncture the pouch.

The tabs can sometimes be spot welded, but they are often made of a material with low enough resistance that spot welding can be difficult. Pouch cells often use aluminum for one tab and coated copper for the other. Both materials can be difficult to spot weld.

You must be incredibly careful when connecting pouch cells because the method of alternately stacking pouch cells results in many cell terminals in close proximity. If two adjacent cell terminals touch when their partners are already welded in series, a short circuit will occur resulting in a flash of light, a loud bang, and probably the disintegration of the terminals. And that's the best case scenario. The worst case scenario would be if the tabs fuse together and cause a short circuit that can't be broken easily, resulting in the cells quickly overheating and potentially exploding. So when you're working with stacked pouch cells, be incredibly careful to avoid short circuits. This usually means that after completing a spot weld between two battery terminals, the terminals should be insulated somehow, usually with heat shrink, electrical tape or kapton tape.

Now we need to also talk about soldering as a method for joining cells. Generally speaking, lithium cells should not be soldered together. I hate to beat a dead horse here, but as we've discussed, heat is the enemy of lithium battery cells. Soldering irons typically operate at around 250°C. What was that thermal runaway temperature of many li-ion cells that we learned about back in Chapter 7? Oh right, 150°C. Lithium cells risk entering thermal runaway at 150°C. So you can see why putting a 250°C soldering iron on the end of a lithium battery cell might be a bit of a problem, right?

To be realistic, unless you're soldering for a really long time, you're probably not going to heat the cell up to its thermal runaway temperature, but you are going to do damage to the cell, and that damage is permanent. Irreparable damage is caused when the cell is heated to high temperatures, and when you solder the cells, you'll cause that damage in small areas that are closest to the cell terminal. It doesn't matter how quickly you can solder them and how short of a period the hot soldering iron tip stays in contact with the cells, it's still going to pass a lot of heat to the cell.

There are basically two issues here: Can you solder on lithium cells, and should you? Technically speaking, yes, you can solder on lithium cells. It will work. But should you? Again, no.

No, no, no.

If you've spent any amount of time researching DIY batteries, then you've inevitably seen people soldering on lithium battery cells. If you search around, you'll also hear people saying that it's ok to solder on lithium cells, that you can do it in a certain way, or that it's fine if you do it quickly, etc. These people are simply misinformed, and are mistaking the fact that while it is possible to solder to lithium cells, it is simply not advisable.

The worst case scenario for soldering is on cylindrical cells such as 18650s, specifically the negative (anode) terminal. As we saw in Chapter 2, cylindrical cells have terminals at either end, and the positive (cathode) terminal is somewhat separated from the contents of the cells by the cap, vent and other components under the positive terminal. However, the negative terminal is the shell of the cell itself and is pressed directly up against the electrolyte filling the center of the cell. This means that when you heat that very thin cell wall at the bottom of the cell, you're directly transferring that heat immediately to the contents of the cell. On the positive end of the cell you've at least got some separation between the terminal and the electrolyte material below, but the negative terminal shows no mercy. Heating the negative terminal with a soldering iron is simply begging for cell damage and premature cell demise.

Pouch cells, with their longer cell tabs, are slightly more resilient when it comes to heating with a soldering iron, but I mean it when I say slightly. Those terminals are very close to the cell's electrolyte and by heating it, you're risking cell damage. Also, many pouch cell tabs can be difficult to solder due to the material used in the tabs. You might have to use special solder and flux depending on the material used. Occasionally you will find pouch cells that have nickel tabs that were spot welded onto the main terminals at the factory. It is much easier (and safer) to solder onto these additional nickel tabs.

If soldering on lithium cells is so bad for them, then why do so many people do it? Well, for one thing, it's a super cheap way to combine cells. You don't need any fancy tools or spot welders or custom mounting brackets. All you need is a soldering iron, some flux and some good solder.

Also, many people build lithium batteries out of salvaged cells that they ripped out of old laptop batteries and other devices. The DIY powerwall community is notorious for this. They are building very large home-energy storage batteries and seek to minimize costs as much as possible. Using salvaged cells and soldering the cells together saves a lot of money. Furthermore, because they are using cells that were basically free or cheap, and because those cells are already of unknown quality and remaining lifespans, they likely consider any cell damage to be inconsequential. What's the problem with losing 10% of your cells' capacities when they were free to begin with?

I see this as both poor craftsmanship and of questionable responsibility in terms of safety. However, people are going to do what they are going to do and my job is simply to give you the facts and information you need so that you can plan your own DIY battery build to fit your needs.

Cell-level fuses

Individual fuses on each cell, also called cell-level fuses, are common on very large batteries and batteries that present a risk to human life and safety. Electric vehicle batteries are a good example. The image to the right shows the simple wire fuses used for each cell in a Tesla electric vehicle. The fuses themselves are small wires attached to the terminal of the battery cell.

The idea behind cell-level fuses is that if any cell in a parallel group should suffer a short circuit, it would overheat and burn the fuse, thus cutting itself off from the circuit. Without a cell-level fuse, a single short circuiting cell could either lead to other short circuits in a chain reaction around it, or slowly drain the rest of the cells down to zero voltage, ruining the rest of the cells in the parallel group.

For the vast majority of hobby-level DIY batteries, cell-level fuses aren't necessary. The chance of a short circuit are very small when using the battery as intended, especially when using good quality cells. However, as the size and thus cost of the battery increases, and as ramifications of failure (such as injury or death) increase, the consequences of a single bad cell bringing down an entire parallel group or pack also increase.

The larger the battery, and the more individual small cells that are in each parallel group, the more useful individual cell-level fuses can be. They are often used in the DIY powerwall community, especially when those packs are made with salvaged cells of unknown histories and qualities. In that case, DIY cell level fuses are often made by cutting the wire legs off of 1/4 W or 1/8 W resistors and soldering them between the cell terminal and the busbar. As we talked about though, soldering is not a great way to connect anything to most lithium battery cells.

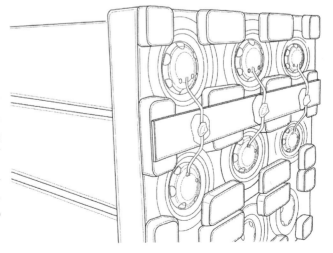

Electric vehicle makers such as Tesla that use many thousands of cells per vehicle have employed sophisticated welding methods to create cell-level fuses. These are often hard to reproduce on hobby-level batteries. One method that has been employed is to cut down the area of nickel strip actually welded to battery terminals to a very small width. This is easier with laser cutting techniques but can be done by hand as well. This creates a short and thin section of nickel that serves as a cell-level fuse.

Suffice it to say that most projects won't require cell-level fuses unless you're planning on driving in it, flying in it, or you think you could stand to lose thousands of dollars if a low quality cell short circuited in it.

Chapter 10: Battery layout and design

Congratulations, we've covered everything we need so far to begin planning our specific battery! This is where the fun starts, because now we can stop talking in theory and start working in reality to plan a battery.

There are many parameters that go into battery layout and design. Some of these parameters are interconnected and changing one will affect the others. It is important to plan for all of our parameters before we can begin choosing cells or even think about starting assembly. Let's tackle the design parameters one at a time.

Voltage

We'll start with voltage, which is one of the most crucial aspects of your battery. The voltage is a measure of the electrical force in your battery. Higher voltage means higher electrical force. Remember what we learned about series wiring in Chapter 6? To increase the voltage of a battery, we'll need more cells in series. In a high voltage battery with many cells in series, those electrons are chomping at the bit, rearing to go! That's due to the higher electron force trying to draw those electrons through the circuit. That also explains why a AA battery (which is 1.5 V) only makes a tiny spark if you short it, but a 9 V battery will produce a much larger spark!

Generally speaking, higher voltage batteries are more efficient, as a higher voltage battery can provide the same amount of power as a lower voltage battery while supplying less current. Fewer amps results in less power being lost as heat.

Higher voltage isn't always better though. You've got to match the voltage to whatever device you are powering. As discussed in Chapter 3, you also need to consider both the voltage drop and the voltage range over the entire discharge curve of your battery.

The voltage drop is the instantaneous reduction in voltage when a load is applied to the battery. The larger the load, the lower the voltage will drop. This is also known as voltage sag. In addition to voltage sag under an instantaneous load, the voltage will also slowly decrease over the discharge curve of a battery. In that case, the voltage will show a larger decrease in the beginning and end of the discharge, and a relatively flatter voltage level in the middle of the discharge curve.

Li-ion cells range from about 3.0 V to 4.2 V, while LiFePO$_4$ cells range from about 2.5 V to 3.65 V. Multiplying those values by the number of cells in series in your battery will give you the voltage range of your battery between its full and empty states.

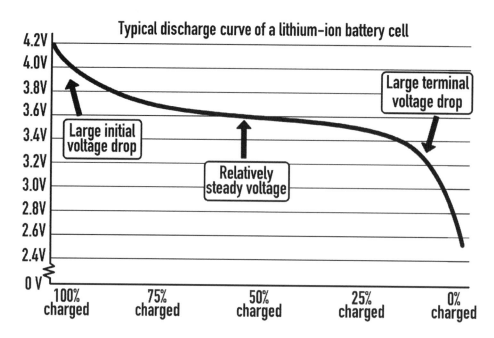

Typical discharge curve of a lithium-ion battery cell

Large initial voltage drop

Relatively steady voltage

Large terminal voltage drop

4.2V
4.0V
3.8V
3.6V
3.4V
3.2V
3.0V
2.8V
2.6V
2.4V
0 V

100% charged   75% charged   50% charged   25% charged   0% charged

If your device requires a constant voltage, you might need to consider using a voltage converter or voltage regulator. A DC-DC converter can take the voltage from your battery and step it up or down to any voltage you specify, assuming it is within the range of that specific converter. DC-DC converters can also be useful if you need a higher voltage but don't want to build a bigger battery with so many cells in series.

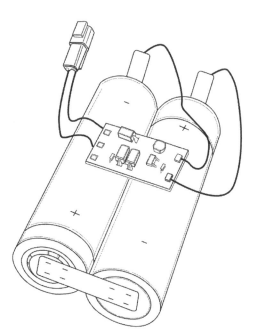

For example, if you need 12 V for a device but only have space (or a weight allowance) for two cells, you can build a 2s1p battery consisting of the two cells in series, and then use a small DC-DC converter to step up the voltage to 12 V. Keep in mind though that using a step-up (or boost) converter will require drawing higher current from the battery to make up for the increased voltage.

Now let's try an exercise to practice designing the right voltage battery for a particular project.

For this example, we'll use an electric bicycle as our project. Let's say the electric bicycle requires a 36 V battery. If we check the controller, we'll find that it actually has a low voltage cutoff (LVC) of 29 V. (This will be different for each electric bicycle and controller; I'm just throwing out a common example.)

That means that we need to make sure that we use enough cells in series to reach the proper voltage. In this case we can use 10 cells in series. A 10s battery would have a nominal voltage of about 37 V (3.7 V per cell × 10 cells in series = 37 V nominal). But we should also make sure that the battery can discharge to a reasonable voltage to meet the low voltage cutoff (LVC). The LVC is 29 V, which would mean 2.9 V per cell on our 10s battery. That's just fine, as lithium ion cells can discharge to 2.5 V, though 2.9 V is of course healthier because it keeps them from reaching their absolute lowest limit. This is also why nearly all 36 V batteries for electric bicycles use a 10s configuration.

What would happen if we tried to use a 9s configuration? That would give us a 33.3 V nominal voltage. A battery of 33.3 V would still work for this application for a little while, but as it discharges, the battery would reach the LVC of the controller at 29 V. For a 9s battery, 29 V would mean each cell is at 3.2 V when the controller reaches the LVC. That still leaves another 20% or so of capacity left in each cell, making this is a bit of a waste. Technically it would be quite healthy for the cells since they wouldn't be discharging very close to their minimum, but it still seems like a waste of good potential, assuming we want to use the entire capacity of our cells.

And what if we made an 11s battery? We'd want to check to make sure that it doesn't exceed the allowable voltage of our device, which is an ebike controller in this case. Let's say that our ebike controller has a type of built-in safety known as a high voltage cutoff (HVC). In this case, if the controller is connected to a battery with a voltage above its HVC, it will shut itself off to protect itself. Let's say that our controller has a 45 V HVC.

An 11s battery would have a fully charged voltage of 46.2 V (11 cells in series × 4.2 V per cell). That 46.2 V is higher than our 45 V HVC. That means an 11s battery would have too high of a voltage for our application in this example. But a 10s battery has a fully charged voltage of just 42 V. That's perfect for this case, so again our 10s battery has the right configuration.

Remember that different lithium cells have different voltages though. The above example used li-ion cells, which have a nominal voltage of around 3.7 V and a fully charged voltage of 4.2 V. LiFePO$_4$ cells have lower voltages. LiFePO$_4$ cells are nominally 3.2 V, fully charge to 3.65 V and can be drained down to 2.5 V. That changes our calculations.

If we rework our example above with the 36 V electric bicycle, a 10s battery using LiFePO$_4$ cells will no longer work. Fully charged, a 10s battery using LiFePO$_4$ cells would only reach 32 V. Considering the voltage sag, the battery would likely reach the low voltage cutoff (LVC) of 29 V in this example almost immediately. Not good. We need more cells in series.

Why don't you try to work out what the proper number of LiFePO$_4$ cells would be for this example with a 36 V battery, assuming a 29 V LVC and a 45 V HVC. Then I'll work it out with you and we'll see if we get the same answer.

...

Ok, have you solved it yet? I hope so, because now it's my turn. Let's see, we need to make sure our lowest voltage is around the LVC of 29 V. If we choose a 12s configuration, then at 29 V we'd see 2.42 V in each cell. That's lower than the 2.5 V we can discharge each cell to. But if we have a BMS protecting our battery then that would be just fine, because the BMS would cut power to the battery as soon as the first cell hit 2.5 V. Now what about the fully charged voltage. At full charge, our 12s battery would give us 43.8 V, calculated as:

Full charge voltage = 12 cells in series × 3.65V per LiFePO$_4$ cell = 43.8 V

That's below our HVC. So this looks pretty good, since our 12s battery stays between the LVC and HVC.

What about an 11s configuration? That would mean that we hit the LVC a bit early at about 2.64 V per cell, and that we leave some charge still in our cells. If your goal is maximum range on this ebike, that would be a problem. But if you are ok with sacrificing some range to give your cells a higher discharge cutoff (and thus a longer life expectancy) then that could be a good thing. At full charge, our 11s battery would be at 40.15 V, which is well under our 45V HVC. So it looks like 11s could also work.

In this case there were two possible answers. An 11s configuration would be a bit more conservative, while a 12s configuration would ensure that you use the full capacity of the cells. You'd definitely need a BMS on the 12s configuration though to make sure you didn't over-discharge the cells, or use another method like a voltage alarm to alert you when the cells are nearly empty.

This is the method you'll use to determine the voltage of your battery in many cases where a fixed range of voltage is required. However, sometimes the voltage isn't critical. In these cases, you will have more freedom.

For example, let's say that we're working with a heating coil, such as a piece of nichrome wire. Maybe we want to make a heated jacket. In this application, we can supply any voltage we want, and it will simply change the amount of heat that we generate from the wire.

Here's an example. Let's say our wire has a resistance of 1 ohm per foot and our length of wire is 10 feet, giving us 10 ohms of resistance. We can use ohm's law of $I = V \div R$ (or current equals voltage divided by resistance) to determine our current requirement and then our heating power. If we used a single 3.7 V li-ion cell, we'd have a current draw of 0.37 A, given by ohm's law:

Current = 3.7 volts ÷ 10 ohms = 0.37 amps

Next we can calculate the heating power in watts. The equation for power is:

Watts = volts × amps

In this example, our heating power equals:

3.7 V × 0.37 A = 1.37 W

If you weren't aware, 1.37 watts is pretty low. That's not going to heat our jacket very much. If we want to increase the power, we can simply increase the voltage. Let's say we used a 10s li-ion battery instead of that 1s battery (which was a single cell). The 10s battery will give us 37 volts nominal. Ohm's law gives us our current of 3.7A which is calculated as:

37 volts ÷ 10 ohms = 3.7 amps

Our new power is about 137 W, calculated as:

37 volts × 3.7 amps = 137 watts

Now we're heating things up! That's probably more heat than we want from our jacket, but you get the idea.

So as you can see, when your voltage range is flexible, your power can change simply by changing your voltage.

Let's do one last quick example and then I promise I'm done talking about voltage for now. When it comes to electric motors, higher voltage means higher speed. DC motors are often rated in terms of KV, which is the number of revolutions per minute (RPM) per volt. So a 100 KV motor would spin 100 RPM at 1 volt or 2,000 RPM at 20 volts.

That same motor would also have a lot more power at higher voltage, but you'd have to be careful not to overvolt it and destroy it, either by burning out its windings or spinning it so fast that it self destructs.

Ok, I think you get it. Enough about voltage. Let's move on to planning for your battery's capacity.

Capacity

The capacity of a battery is highly related to the maximum continuous discharge current rating of the battery. But we'll start with capacity as it's a bit simpler, and then we'll work our way over to current.

The capacity of a lithium battery is kind of like the size of a gas tank in a car. The larger the gas tank in a car, the longer it will drive. And the more capacity in your battery, the longer it will work.

The capacity of a lithium battery is usually measured in amp hours (Ah) or watt hours (Wh). Amp hours are a measure of the number of amps the battery can supply for one hour. Watt hours are similar, but measures the number of watts the battery can supply for one hour.

Between the two, amp hours are more common in the industry. However, watt hours actually give a better indication of the total energy in a battery. That's because watt hours are calculated by multiplying the battery's nominal voltage by the Ah rating of the pack. So a 36V 10Ah battery would be approximately 360 Wh.

While the amp hour rating is a specific measurement for each battery, watt hours also consider the voltage of the battery and is thus a measurement of the total energy in a battery. That makes watt hours better for comparing between batteries of different voltages. A 12V 10Ah battery and a 24V 10Ah battery both have the same amp hour rating, but the 24V battery has twice the watt hours (240 Wh vs 120 Wh) because it stores twice the amount of energy. For batteries or battery cells of the same voltage, Ah is usually used for capacity comparison.

As we learned in Chapter 6, the capacity of a lithium battery is increased by adding more cells in parallel (or using larger capacity cells). So if we needed a 10 Ah battery and we were using 2.5 Ah cells, we would want to put four cells in parallel. That gives us a total capacity of 10 Ah by:

2.5 Ah × 4 cells = 10 Ah

We could have also just used a single 10 Ah cell, but those large cells come in fewer options and are usually limited to pouch and prismatic cells. Using smaller cells in parallel gives us more options.

The exact capacity of your battery will of course depend on your specific applications. If you're building a small drone, then you might be fine with a 4 Ah battery. An electric bicycle usually uses a battery of between 10 Ah - 20 Ah. An electric vehicle's battery is usually over 100 Ah and sometimes can be many hundreds of amp hours!

To calculate how much capacity your battery will need, it is helpful to first think in terms of watt hours. Let's use a simple example first. Let's say your load is a light bulb which takes 12 V. Let's also say that our example light bulb draws 1 A continuously during use. If your application requires that your battery should power the light bulb for 8 hours, then you can calculate how much energy you'd need.

First, calculate the watts used by the light bulb. Remember: watts = volts × amps

That means that the watts used by the light bulb are equal to 12 watts, calculated as:

12 volts × 1 amp = 12 watts

This gives us 12 watts of continuous power being used by the light bulb. Remember, watts are a unit of power, not energy.

Now we need to determine the amount of energy used by the light bulb, which is where watt hours are used. A watt hour (Wh) is a measurement of energy, not power. To determine the amount of energy our light bulb uses, we just multiply the power it uses, in watts, by the amount of time it uses that power, in hours.

Watt hours consumed = continuous power (in watts) × time (in hours)

In our case we want the light bulb to run continuously for 8 hours. To calculate the total energy needed for 8 hours, we multiply 12 watts by 8 hours. This gives us 96 watt hours.

Ok, now we're getting somewhere. Our light bulb needs 96 watt hours to operate for 8 hours. If we're using a 12 V battery, we just need to find the right number of amp hours that will give us

96 watt hours of energy in the battery. The watt hours of a battery are calculated by multiplying the voltage of the battery by the amp hours.

Watt hours of a battery = voltage × amp hours

That means that to get the amp hours of a battery, we would take the watt hours of the battery and divide it by the voltage.

Amp hours of a battery = watt hours ÷ voltage

In our case, that means we divide the watt hours needed (96 Wh) by the voltage of our battery (12 V). This gives us 8 Ah.

96 Wh ÷ 12V = 8 Ah

That means that for our example of a 12 V light bulb that requires 1 A of current and operates for 8 hours, we need a 12V 8Ah battery. This should make sense, since we already knew that the bulb draws 1 A, which means that every hour it draws 1 A per hour, or 1 Ah. So for 8 hours of use, we'd need a total of 8 Ah.

Let's try another example. Let's say we have a 100 W heater that operates at 24 V. And let's say we want to run that heater for 20 hours. We'd want a 24 V battery that can provide 2,000 watt hours of energy, calculated as:

100 W × 20 hours = 2,000 watt hours

To determine the Ah of the battery we'd need, we simply divide 2,000 Wh by 24 V.

2,000 Wh ÷ 24V = 83.33 Ah.

That means we'd need an approximately 24V 83Ah battery to run that 100 W heater for 20 hours.

Now I want you to try a couple of examples, just to make sure you're totally on board with the process. Try these three problems:

1) We have a load that requires 36 V and draws 5.5 A. We need it to run continuously for 5 hours. What size battery do we need?

2) We have a 48V 10Ah battery. If we connect it to a load that draws 25 A continuously, how long will the battery last?

3) Alright, this one is a bit trickier, but I think you can handle it. Take that same 48V 10Ah battery from question 2 above. Let's first connect it to a 50 W load for 2 hours. Then we'll disconnect it from the 50 W load and connect it to a 20 W load instead. How long will it last on the 20 W load until the battery is depleted?

Answers:

1) This load is 36 V × 5.5 A = 198 watts.
At 5 hours, it uses 198 W × 5 hours = 990 Wh.
990 Wh ÷ 36V = 27.5 Ah
So we need a 36V 27.5Ah battery.

Alternatively, you could have gotten the same answer with a short cut. Just multiply the 5.5 A it uses by 5 hours, giving you 27.5 Ah.

2) We know that our load is larger than our capacity rating (25 A is larger than the 10 Ah rating) so we already know this battery will last less than 1 hour. Dividing the 10 Ah capacity by the 25 A load gives us 10 Ah ÷ 25 A = 0.4 hours, or 24 minutes.

3) First let's calculate the total energy of the pack in watt hours. 48 V × 10 Ah = 480 Wh. If we connect a 50 W load for 2 hours, that is a total of 100 Wh that we will drain from the pack (50 W × 2 hours = 100 Wh). So after 2 hours of supplying 50 W continuously, our 48 V battery is down to 380 Wh (480 Wh - 100 Wh = 380 Wh). Now we connect it to a 20 W load. Dividing the remaining 380 Wh by 20 W gives us 19 hours (380 Wh ÷ 20 W = 19 hours).

If you got all those answers correct, great! If you missed one or more, make sure you go over the solutions until you understand where you made a mistake. The math isn't too hard, but it's very important to go slow and make sure you get this right. If you make an error in planning your battery, you could end up with not enough capacity to fulfill your needs, or too much capacity. (And that would give you a battery that is larger and more expensive than you needed!)

One last note: remember that cells often don't provide their entire capacity even when new, and that their capacity will decrease as they age. If your project requires 10 Ah of capacity and you use cells that add up to exactly 10 Ah, it wouldn't be surprising for your battery to only provide 9.7 Ah when new, and closer to 9.0 Ah after a few hundred charge cycles. So if your project allows for it, consider building a battery that has slightly more capacity than you need.

Maximum continuous current

When introducing capacity in the previous subsection, I mentioned that the capacity is highly related to the maximum continuous discharge current. This is because the larger the capacity of the assembled battery, the larger the maximum continuous discharge current of the assembled battery will be. That is why I spent so much time in Chapter 5 discussing how to calculate C rates. (And that's why I forced you to take a mini test on C rates too!)

Even if you use the exact same battery cells to build two batteries of different Ah capacities, the larger capacity battery will be capable of a higher maximum discharge rating. A higher maximum discharge rating is the same thing as a higher continuous current rating.

This is where C rates come into play. Let's take a common 18650 cell for example: the Panasonic/Sanyo NCR18650GA. This cell has a capacity of 3.5 Ah and is rated for 10 A maximum continuous discharge rate. To determine the C rate of this cell, we divide the maximum continuous current that it can supply by the capacity in amp hours, giving us:

10 A ÷ 3.5 Ah = 2.86 C

That means that not only is the cell rated for 2.86 C maximum continuous discharge, but any battery pack built out of these cells will also be rated for 2.86 C maximum continuous discharge, no matter how big the battery pack is. However, a larger battery pack will be able to supply a larger current.

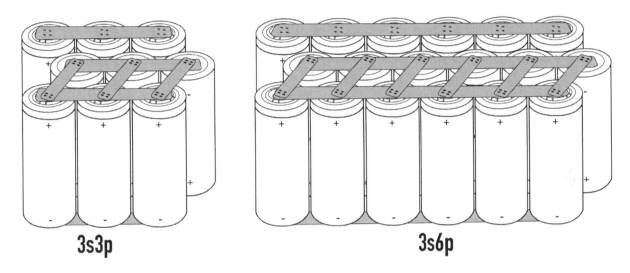

3s3p　　　　　　　　　　**3s6p**

Let's take a look at two potential battery packs built from these cells. We'll make a hypothetical 3s3p pack and a 3s6p pack, shown above. Now let's compare them. The first battery pack will be 11.1 V nominal (3 cells in series × 3.7 V nominal per cell) and 10.5 Ah (3 cells in parallel × 3.5 Ah per cell). The second battery pack will also be 11.1 V nominal, but it will be 21 Ah, or twice the capacity of the first battery pack. Both of these battery packs will be capable of a 2.86 C discharge rate. However, that 2.86 C discharge rate equals different absolute currents for the two different battery packs. The first battery can supply 30 A continuously, while the second battery can supply 60 A continuously.

How do we determine the maximum continuous discharge current that the pack can supply? There are two methods. The first method is to multiply the maximum continuous discharge current of the cells by the number of cells in parallel. For the first battery that would give us:

10 A maximum continuous discharge × 3 cells in parallel = 30 A discharge rate

The second method is to multiply the C rate of the battery pack (or of the cells, since it's the same for both) by the capacity of the battery pack in amp hours. For the first battery that would give us:

2.86 C × 10.5 Ah = 30 A discharge rate

It is important to design your battery so that it can provide enough current for your application. Ideally you want it to be able to provide more current than your application requires. In engineering terms, this is known as a factor of safety.

Factor of safety = maximum allowable load ÷ actual load

If I'm trying to power a light bulb that draws 5 A continuously, the lowest maximum continuous discharge rate my battery can safely have is 5 A. That would give me just enough current to power the 5 A light bulb. If my battery is rated for 10 A maximum continuous discharge instead of 5 A, then I'll have a factor of safety of 2, because my battery can provide twice the current that the light bulb requires. If my battery is rated for 20 A maximum continuous discharge, then I'll have a factor of safety of 4, because my battery can provide 4 times the current needed by the load.

Generally speaking, you want to have some level of factor of safety. It doesn't need to be as high as 2 or 4, but it definitely needs to be higher than 1. A factor of safety of 1 means that your battery is working at its absolute limit, providing the exact maximum amount of power that it is rated for. A factor of safety of less than 1 means your battery isn't even strong enough for the load it is powering. That's obviously unacceptable.

The reason a factor of safety is a good idea is because most lithium battery cells suffer a decrease in performance as you approach their maximum rated values. The 18650GA cells we used in the example above are rated at 10 A, but they get very hot at a constant 10 A discharge and don't provide their full 3.5 Ah at that level of discharge. Instead, they give closer to 3.2 Ah when pushed to their max. I like to keep the current in 18650GA cells below 7 A. The factor of safety for a 10 A capable cell operating at just 7 A can be calculated as:

Factor of safety = 10 A ÷ 7 A = 1.4

Most medium to low power projects can be handled by $LiFePO_4$ and li-ion cells. If your project requires a really high maximum discharge current though, you'll either need to place many cells in parallel to supply enough current, or you'll need to use RC lipo cells. Remember, RC lipo cells are the dangerous ones that you need to be extremely careful with. Sometimes though, they are the only option. A prime example is when the highest levels of power are needed and weight or space is an issue.

If you have a project that needs 100 A and you only have room for a few 3 Ah cells, you're not going to find a standard li-ion or a $LiFePO_4$ cell that will work. You'll have to use some high power RC lipo cells. However, if you had the space and weight allowances, you could just parallel many li-ion or $LiFePO_4$ cells and reach that 100 A discharge capacity. Assuming those cells were rated at 5 A maximum continuous discharge, you'd need 20 cells in parallel. For cells with a 10 A maximum continuous discharge rating, you'd only need 10 cells in parallel.

As you can see, there are nearly an infinite number of ways to design most battery packs for different needs. Voltage, capacity and current all affect each other as well as the final specifications of a battery. By making compromises or changing the balance between those three parameters, a myriad of options is possible.

Choosing appropriate cells

It's nearly impossible to choose the right cells for your project without knowing your final voltage, capacity and current needs. Once you know those, you can go ahead and find the right cells to fit your needs.

Current rating

The first aspect of choosing the right cell for most projects will likely be the maximum current rating. If you have a 40 A load, then you'll need enough cells in parallel to supply 40 A. If you know that you only have room (or the budget) for 4 cells in parallel, then you better make sure each of those cells can provide at least 10 A continuously.

Using that same example with a 40 A load, if your space and weight requirements allow it, you could also build an 8p battery instead of a 4p battery. In that case, your cells would need to be capable of supplying at least 5 A each.

Now what if you need that 40 A load to run for 30 minutes? It's pretty simple to calculate the capacity you'll need. Just remember to convert your minutes into hours or you'll be calculating the number of amp minutes that you'll need.

Required capacity = 40 A × 0.5 hours = 20 Ah

Your battery should have a capacity of at least 20 Ah to run that 40 A load for a half hour. That means that however many cells you use in parallel, they need to add up 20 Ah.

Let's say we're using 3.5 Ah cells that can provide 10 A continuously. A 4p battery will gives us the 40 A continuous discharge rate that we need, but it will only give us 14 Ah of capacity. That's not enough to run our load for the half hour we used in the previous example. In fact, it will only give us about 21 minutes of run time.

Runtime = 14 Ah ÷ 40 A = 0.35 hours or 21 minutes

Since our capacity doesn't give us enough run time, we'll need more cells in parallel. A 6p configuration will give us 21 Ah if we're using those same 3.5 Ah cells. That is above our requirement of 20 Ah, perfect! It will also give us a maximum discharge current of 60 A, when we need at least 40 A of continuous current. That will give us a factor of safety of 1.5 for the maximum current draw, which is great!

Now we've worked out a number of examples and I hope you see how capacity and maximum current draw are integrally related when it comes to building a battery pack and choosing cells. Capacity and maximum discharge current are two of the most important factors, but there are a few other factors that may be relevant to your project. These can also affect which cells you choose.

Cycle life

Cycle life varies from one type of cell to the next. Some cells can only perform 200 cycles. Others can last for more than 2,000 cycles. Depending on the needs of your project, cycle life could be a limiting factor.

If you need the highest cycle life for your project, you're probably stuck with LiFePO4 cells. They are the most widely available cells that can perform over 1,000 charge cycles.

If you can replace the battery whenever you want, then cycle life might not be very important for you.

Weight

Weight is another issue. In Chapter 3 we talked about how LiFePO4 cells are the heaviest cells and li-ion cells are the lightest. Keep that in mind if weight is important.

Size

Just like weight, physical size can be a limiting factor for many projects. To cram the most energy into the smallest space, the NCA chemistry of li-ion cells are what you need. Remember though that they have somewhat lower power, which is the tradeoff for being so energy dense.

LiFePO4 cells are the largest cells you can use. If you have very limited space, you're unlikely to use LiFePO4 unless its benefits including safety and long cycle life are critical for your project.

Safety

Speaking of the benefits of LiFePO4, if your project requires the highest level of safety standards, LiFePO4 is a great option. As we learned, LiFePO4 is the safest chemistry due to its incredibly high temperature needed to cause thermal runaway as well as its slow and difficult combustion.

Li-cobalt and RC lipo, on the other hand, are the most dangerous lithium chemistries available. If safety is a concern, those are probably not the cells you want to use. If you're stuck with them due to other limitations, make sure to use a BMS to protect the cells during both charging and discharging. A BMS can make those cells much safer. Used properly, li-cobalt and RC lipo can be quite safe. But in comparison, they are still more dangerous than other chemistries of lithium batteries and accidents with those cells are generally more catastrophic.

Grouping cells in a battery

Once you know the required voltage, capacity and maximum discharge current needed by your battery, and you've chosen your cells, you can begin designing the layout of your battery. The method you use will largely depends on the type of cells you choose.

If you're using prismatic or pouch cells, you're more than likely going to use a stacked setup. This will mean that you're placing the cells in rows or stacks with alternative directions to align the positive terminals of each group with the negative terminals of the next group.

If you're using 1p parallel groups, meaning just one cell, this is very simple. Just stack or line up your cells with alternating cathodes and anodes, then connect each one to the next in series.

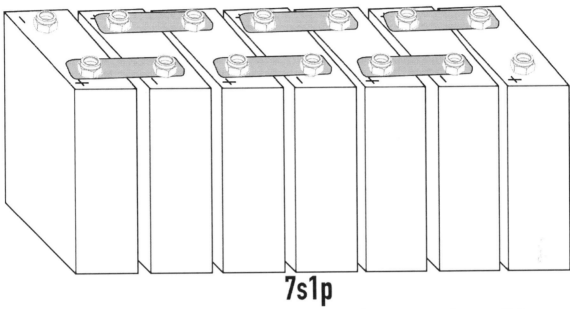

7s1p

If you're using multiple cells in parallel, you may want to first connect the parallel groups. After you've stacked and connected the parallel groups, you can stack the combined parallel groups and make your series connections. The order (parallel groups first versus all connections at once) will largely depend on the type of connections you're making and the shape of the battery. Sometimes it is easier to work with parallel connections first, but other times it doesn't make a difference. With busbars on linear battery cells, it might just be easier to lay down the busbars at once, assuming they are long enough to do the parallel and series connections at once.

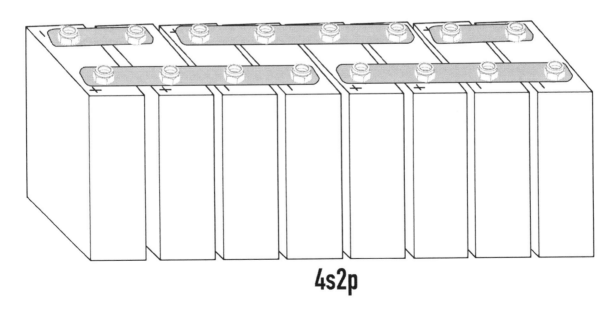

4s2p

It's hard to get too creative with pouch cells or prismatic cells. Their shapes simply don't allow for many variations in packing. That's why the typical stacks or rows method is almost always used with pouch and prismatic cells.

Cylindrical cells will give you much more freedom and creativity in battery shape and layout. With cylindrical cells, there's almost no limit to the ways you can build a pack. As long as you can hold the cells together using any of the methods we discussed in Chapter 9, you can build a pack in that shape.

With cylindrical cells, there are two main stacking orientations that can be used, linear packing and offset packing. Linear packing is less efficient because it leaves larger voids in the center of the battery pack. Offset packing is the most efficient method.

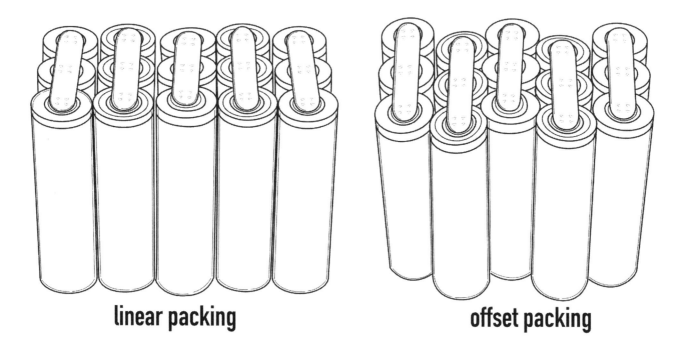

linear packing   **offset packing**

Linear packing is common when using snap-together battery holder blocks. Most of these blocks are only designed to snap together using linear packing. Some molded plastic battery holders use offset packing, but they are generally less modular and come in pre-made sizes. This gives you less freedom than single snap-together blocks.

Offset packing is more common when using the hot glue method to hold cells together. Not only does it take advantage of the space savings of hot glue versus rigid cell holders but it also gives two points of contact on every cell on which the hot glue can affix itself.

Regardless of the method of alignment you use in your cylindrical cell battery, you'll want to pay close attention to where you'll make your series and parallel connections. Remember that in Chapter 9 we discussed how much current each nickel strip can carry. Whatever conductor you're using to make your series connections, if it can't carry the entire current that the battery can supply then you'll need to use multiple series connections.

For example, look at this diagram of joining two 4p parallel groups that are connected in series.

A single piece of nickel in the series connection limits the current carrying ability

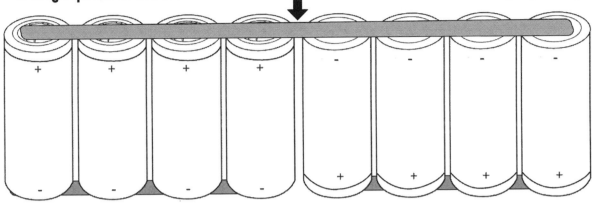

If we use a single piece of 0.15 mm thick and 8 mm wide nickel strip to join these in series at one point, we can only draw approximately 5 A from this battery. However, if we used four strips of nickel, we could draw 20 A from the battery (assuming the cells are rated for at least 5 A each).

Several layers of nickel in the series connection increases the current carrying ability

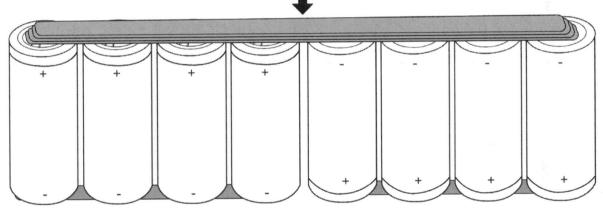

If we put all of these eight cells in a straight line, we'd need to stack our nickel strips up on top of each other. We could use four nickel strips long enough to span all eight cells, but that would be a bit of a waste. The current flowing through the nickel between the first and last cells in the eight cell line don't see as much current as the nickel in the middle of the eight cells, which carries the full 20 A. That means we can actually save on material by building a pyramid of four nickel strips that are each progressively shorter by one cell width on each side.

Layers of nickel in a series connection can be stacked in a pyramid to save material

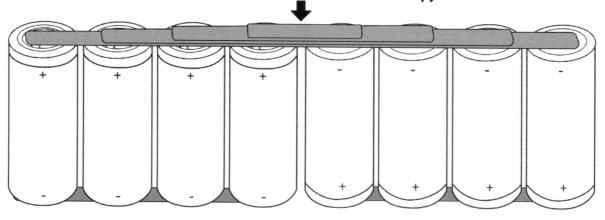

That pyramid structure is the proper way to join cells in series with nickel strip when the cells must be in a single straight line. But what if the cells in series aren't required to be a single straight line? In that case we can be even more efficient. If we align the same cells in two parallel straight lines of four cells each, we can use a single layer of nickel connected between each parallel group of cells in the 4p groups.

Ideal layout maximizes cell-to-cell series connections

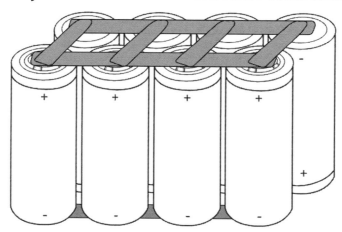

Now we're creating four individual series connections instead of four series connections that are all stacked on top of each other. This still gives us a 20 A discharge capability because the current has four pieces of nickel strip to flow through, with each piece of nickel capable of supporting 5 A. This method uses less nickel and also requires fewer welds at each location than the pyramid stacking method.

When possible, it's always better to align series connections so that there can be as many individual cell-to-cell connections as possible. This is better than stacking multiple nickel strips on top of each other to beef up the connection. Both methods are valid and will produce the same equivalent electric circuit, but increasing the cell-to-cell connections is simply better form.

Let's look at an example of an 8s4p pack. We could build a simple rectangle of cells that is eight cells long and four cells wide. This would give us two options for designing our connections.

The first option would be with two parallel groups per row. The second option would be with one parallel group per column. Both would give us the same circuit, but the second option would allow us to make more efficient series connections.

Non-ideal layout of 8s4p battery (Side 1 requires four single-point series connections)

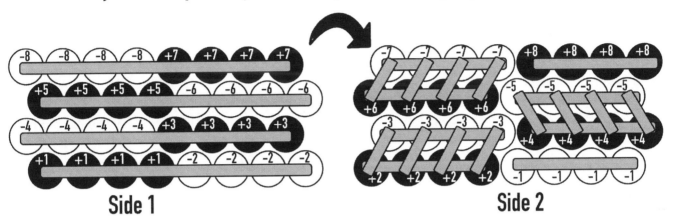

Side 1   Side 2

Ideal layout of 8s4p battery (both sides have maximum cell-to-cell series connections)

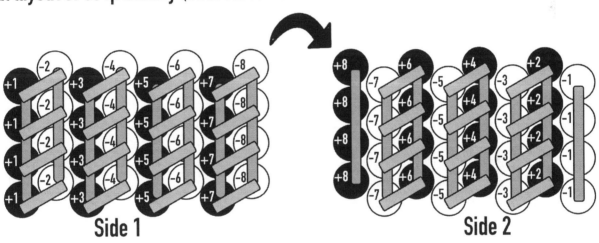

Side 1   Side 2

This same concept also applies when producing other geometric or abstract shapes. You always want to maximize the number of cell-to-cell series connections when possible.

The more cells you have in your battery, the more complicated it will be to plan the connections. One of the most confusing parts of planning the inter-cell connections when using cylindrical cells is that the connections are different on each side of the battery. And it's important that they're different - if you make the same connection between two cells on both sides of a pack then you've just shorted the cells to each other!

One trick I often use is to draw my battery cells out on a thin piece of paper using a heavy pencil or marker. This makes it easy to see marks made on the other side of the paper, especially when held up to the light. If I draw both sides of my battery on opposites sides of the paper so that they align, I can flip the paper back and forth as if I'm physically flipping my battery over. Then I can hold the paper up and "see" the connections on the other side of the battery. If I have

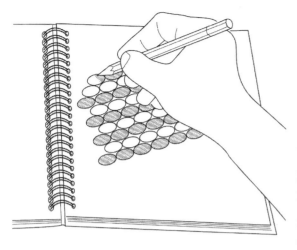

connections between the same two cells on both sides of the paper, I know I've royally messed up. That would be a disastrous short circuit.

Using paper to design your battery helps you avoid these connection mistakes. You can also layout your battery using a computer program. I generally use a Photoshop knockoff, but even the simplest drawing programs will work. You can draw just one side of your battery and use two different colors to represent the connections on opposite sides of the battery. Call me old fashioned, but I still prefer to flip my paper over to draw each side.

If you really want to use a computer to design your cell layout but still like the idea of the physical paper diagram (it really helps during the actual welding/connecting phase!) then you can always design both sides on the computer, print them out and tape them together back-to-back. Then when you are making your connections, you can easily flip the paper over to each side and see a representation of how your battery should look. It's a little more arts and craftsy, but it gets the job done.

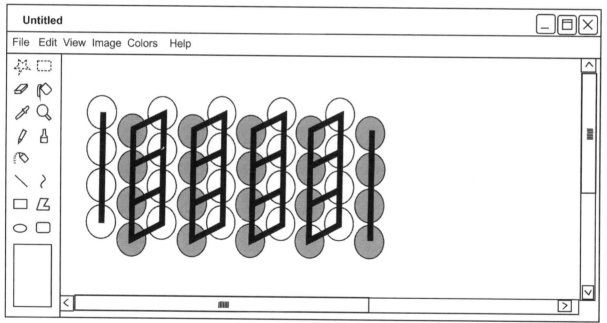

Making different battery shapes

No matter what type of cell you're using, the easiest shape of battery pack to build is a rectangle. Easiest by far. That doesn't mean rectangles are the only shape you can work with, but we'll at least start from there.

As mentioned previously, pouch cells and prismatic cells are pretty difficult to build abstract shapes with, but they lend themselves really well to rectangular shapes. And fortunately for us, rectangular shapes work quite well in many applications. Everything from electric skateboards to home energy batteries and electric vehicle batteries usually come in rectangular prisms.

Why? Because it's easy, mostly. Why go to the trouble of designing a fancy shaped home battery when a square easily mounts in a closet or on a wall. Homes are built with 90 degree angles, and so are the boxes that we'll probably use to enclose our batteries.

Same things with cars. Things are usually relatively flat and angular in most areas of the trunk and undercarriage. Why? It's cheaper and simpler. Fancy curves belong on the outside where people see them. Storage places - the exact places you're going to stick your batteries - those are usually flat with corners. Same thing with electric skateboards. They've got a flat deck, so why wouldn't they want a flat battery?

The fancy shapes usually come about when you're trying to put batteries somewhere they weren't normally intended to go, or you're trying to maximize available space. Or both.

Electric bicycles are a great example. Triangle shapes are often used in ebikes because they fit into the triangular shape of the front of a standard rigid bicycle frame. Bicycles with suspension have even weirder shaped frames and often require even weirder shaped batteries.

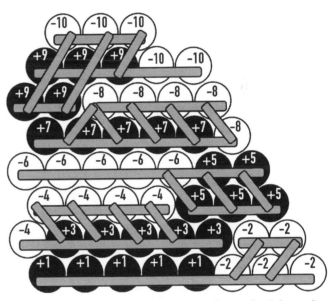

10s5p triangle battery for electric bicycle

If you're trying to conceal batteries inside of odd shaped enclosures, then you'll also run into weird shapes that can be a bit more difficult to construct.

In these cases, if you already have the physical thing that will contain your battery (like a bicycle frame or an electronics enclosure, then the best plan is to make a template. Measure your shape and then draw it to scale in a 2D computer program like Paint or Photoshop. Now you can play around with all sorts of interesting shapes of cells. When you want to test your design, simply print your drawing, cut it out with scissors and test fit it into your device.

I once built an electric bicycle battery for a customer's odd shaped bicycle frame using this exact method. From halfway around the world he emailed me the measurements of his frame. I then drew up a template for the cells and emailed it back to him. He test fit the paper template into his frame to ensure it would fit. I then built the battery to fit that template and shipped it 7,000 miles to him. It fit in the bike frame like a glove.

Once you have your shape all planned out, the hard part isn't quite over. Extremely odd shaped batteries can be difficult to connect, especially if there are large gaps or voids in the battery. If you're using pouch or prismatic cells, you can use flexible wires to span long gaps between cells.

Just make sure to use a thick enough wire gauge to account for the amount of current that will be flowing over the distance. If you're using spot welded cells, spanning gaps can be trickier, but it's still possible.

For really abstract shapes, a spot welder with handheld welding probes will make the battery construction much easier. The stubby electrodes on most hobby-level welders can be very limiting. Most only reach about two cylindrical cells deep. By bending the copper electrodes out and tilting the battery during welding, you might be able to get as deep as four cells. But if you use a welder with handheld probes, you can reach anywhere in the battery and not be limited by the reach of the stubby arms on the front of the welder.

Remember what we talked about in the last section regarding making sure sufficient current can flow between series-connected cells? In abstract shaped batteries, you'll need to be extra careful that you're using enough layers of nickel strip to carry sufficient current between groups welded in series. This is easier in rectangular batteries because series connections can often be made between every cell in a group. If you've only got enough room to fit a single connection between two cells in two adjacent parallel groups, you'll have to stack enough nickel to have a sufficient current capacity in that connection.

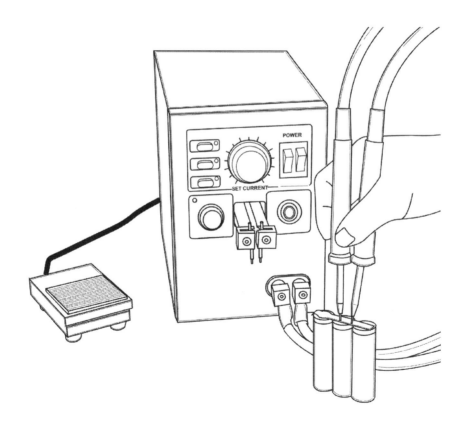

Chapter 11: Making battery connections

Here we go, the time has come and we're ready to build an actual battery! But before we get too carried away, we need to get a few things ready. First and foremost, we need to ensure that we've prepared a safe working environment.

Safety considerations

We discussed the importance of safety in general when working with lithium batteries back in Chapter 7, so I won't harp on it too long here. It's important enough for a quick review though. These are the main safety tips to remember when you're ready to actually start building a battery.

Always remove jewelry and wear gloves. A wedding ring makes a great short circuit, as do sweaty palms.

Clean the work area thoroughly before laying out your batteries. If possible, lay out some large sheets of paper on your work surface. They'll cover any existing metal debris you missed and make it easier to see new bits and pieces laying around. Avoid newspaper as the black and white print can make it hard to see small objects. I like white butcher paper. Even a few sheets of computer paper can work well too - just tape them together at the edges so you don't lose things under them.

Wear safety glasses if you'll be spot welding or soldering. Spot welders can throw sparks. Soldering on anything springy like a wire or nickel strips can accidentally fling molten solder around if it springs back while soldering. In a competition between your eyeballs and sparks or molten metal, your eyes lose every time. You've only got two of them. And if you value your depth perception, having only two eyes leaves no redundancy.

Make sure you don't have direct airflow on your battery. If you're soldering, you probably want a fan to help remove the fumes, but point it up above where you're working. The last thing you want is for anything to catch in the moving air and knock some metal onto your exposed battery contacts. It sounds far fetched, but you'll probably have a pile of wires or nickel strips laying out on the big sheet of paper I just told you to use. Fan, meet sail.

Lastly, use common sense about everything you do. Double check before you make a connection to ensure that it is the correct connection to make. Don't juggle metal tools over your exposed battery. Don't leave your exposed battery sitting around in a highly trafficked area. On that note, if you stop for the day before you finish a battery build, cover it with a plastic bag or sheets of paper to ensure nothing falls on your battery and shorts it. When I have to leave a partially completed battery laying out unattended, I place a big "danger" sign on it to make sure no one messes with it.

Basically, use your head when it comes to safety during your preparations and you should be fine.

One safety note specifically about cylindrical cells - Most cylindrical cells such as 18650 cells are constructed with the entirety of the bottom and the sides of the cell connected to the negative terminal of the battery. The sides of the shell even wrap around the top, meaning the negative and positive terminals of the cell are separated by just a few millimeters at the top of the cell.

The cells have heat shrink around their outer shell to protect the cells from shorting against each other. If that heat shrink was ever damaged, such as by constant vibration or chafing, there could be the potential for a short circuit. This is most plausible on the top of the cell where connections between the cell could rub against the heat shrink. This isn't particularly common, but cases where nickel strip welded to the top of the cell has vibrated and subsequently worn away the heat shrink have been observed. Such cases can often lead to catastrophic failure by shorting one cell, leading to thermal runaway of the entire pack.

Most reputable cell manufacturers include an insulating paper or plastic disk or washer under the heat shrink on the top of the cell to add an extra layer of protection against short circuits occurring on the top of the cell. This is usually a white circle that can be seen at the top of the cell, just under the heat shrink. However, cheap cells often lack this second layer of protection.

Insulating paper washers made from sticker paper are commercially available and can solve this problem. They are sometimes referred to as "fish paper". The paper gaskets can be applied on the top of a cylindrical cell to provide an extra layer of short circuit protection and won't interfere with connections. For good quality cells that already come with an extra insulating gasket, adding this insulating sticker would be a second redundancy against shorts, but not a bad idea. For cells without an insulating gasket included by the manufacturer, this insulating sticker would be a good second line of defense.

One last thing to mention. You probably already did this when you checked your cells, but you'll want to make sure all cells are at the same voltage before you begin your connections. It's more important for parallel cells, as the instant you make the parallel connection, cells of different voltages will try to balance each other. If the voltage difference is large, a huge amount of current will flow very quickly through your parallel connections.

This is also important for series connected cells because any large voltage imbalance will have to get balanced eventually by either a BMS or balance charger. The larger the initial voltage imbalance, the longer it will take to get your cells balanced during their first charge.

Cell matching

As previously mentioned, when you combine your cells into larger battery packs, you want to make sure that your cells all have the same voltage and the same capacity. If you are using new cells, this is easy as testing them with a voltmeter to ensure they are at the same voltage. If you're using recycled or salvaged cells, this is a much more complicated process.

Salvaged cells are not only going to be at different charge levels, but they'll also have different capacities. Just because two cells are both at 3.5 V doesn't mean they have the same capacity. One of them could have been a 2.5 Ah cell and the other a 3.5 Ah cell. Or maybe they were both originally 3.0 Ah cells, but one has been used for many more cycles and is now down to 2.8 Ah.

When you're using salvaged battery cells, you'll want to test each cell individually to determine its capacity. There are many different lithium battery cell testers out there. Most are for 18650's but there are some bare board testers that can be used on most lithium cells.

After testing each cell, it is helpful to write the capacity on each one with a felt tip marker so you don't mix them up. You'll want to make sure that you group together like-capacity cells when you create parallel groups.

For example, let's say you are making a 7.4V 20Ah battery pack. You might have a total of 18 salvaged cells, with eight cells that are all around 2.5 Ah and ten cells that are all around 2 Ah. In this case, instead of putting nice cells in each parallel group, it would be better to group the eight 2.5 Ah cells together in one group and the ten 2 Ah cells together in the other group. That way each parallel group is approximately 20 Ah. Just remember that your smallest group (by number of cells) will determine how much current you can draw from the battery. If you want to limit your cells to 2 A each, then you could draw a maximum of 16 A from this battery because your smallest parallel group has only 8 cells.

You can see how this can get complicated as the number of batteries increases. That's another great reason why it's better to start with brand new cells instead of using mystery cells or salvaged cells. With brand new cells you will never have to play this matching game. All you have to do is double check that all the voltages are the same and then start building groups with the same number of cells in each group.

Cell alignment and containment

Before you can begin to connect your cells, you need to have them already aligned and contained in whichever method best suits your project. We discussed many different types of cell holders and enclosures in Chapter 9.

Once you begin connecting your battery cells, the amount of potential energy grows quickly. The last thing you want to be doing is be moving around large connected cell groups unnecessarily. Always organize and layout your cells in the proper orientation before you begin making your electrical connections.

Once you've begun your connections, you'll want to minimize the amount of battery movement. This is for both safety and quality reasons. The more you move your battery, the more chances you have for accidents. Additionally, the more rigid your connections are, the more likely they are to fatigue or loosen as your move your battery around and stress those connections.

Occasionally you won't be able to lay out your battery completely before the assembly process begins. This is often true for very large batteries that must be first built into sub modules and then modules that are ultimately combined into a single large battery pack. Other times the battery will not have a self containing enclosure, such as drone batteries that are simply enclosed in a layer of heat shrink wrap after the connections are made. In these cases, use your best judgment to ensure that your battery is handled safely during the construction process.

Bolted connections

Bolted connections are generally the easiest method to join your cells together, but pretty much limit you to prismatic cells or Headway's unique cylindrical cells.

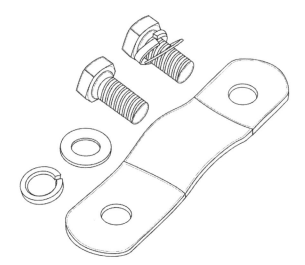

For bolted connections, you can use a variety of conductive materials to bridge the battery terminals and make your connections. The two most common options are busbars and flexible wire.

Busbars are usually made out of either aluminum or copper. Both materials have similar resistances and work well for electrical connections. Copper is a slightly better conductor, but it's also more expensive and corrodes more easily.

It's best to try and match the thermal expansion of whatever material you're using as closely as possible. As the connections heat up, dissimilar metals will expand at different rates which can loosen the connections over many heating and cooling cycles. Lock washers, especially split washers, can help prevent this loosening phenomenon.

For wire connections, you'll probably end up crimping or soldering a ring connector onto the end of your wires. You can simply wrap the wire around the bolt or threaded axle and clamp it between washers as well, but this method has a few disadvantages. It will likely result in only part of the wire making contact with the cell terminal and will reduce the amount of current you

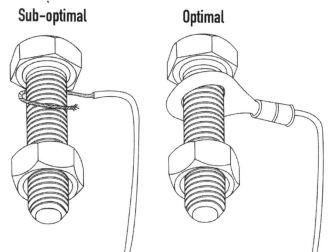

Sub-optimal   **Optimal**

can carry through the connection. If your battery will experience vibration or movement, this method can also result in the wire slowly working itself free.

Ring connectors are a better way to mount wires to threaded axles or bolts. It is best to crimp them as opposed to soldering, as the solder can weaken the wire. This is more important for batteries that will experience motion and vibration. If your battery will be stationary, soldering your connectors is probably fine.

If you're using wire to make the connections between cells, be sure to use a sufficiently large gauge wire to carry the current you need. The appropriate gauge wire is different for every project, and you'll need to determine your specific requirements. Different wire materials, whether copper or aluminum, are rated for different levels of current at different thicknesses and under different atmospheric conditions (i.e. in a vacuum, in ambient air or in flowing air). Consult the appropriate charts for your specific wire type to calculate the proper wire size for your project. A helpful table on choosing wire sizes can be found at http://www.engineeringtoolbox.com/wire-gauges-d_419.html.

This is of course also true for the wires that will handle the charging and discharging of your battery. Any wire that will carry the full current of your battery should be appropriately sized based on the conditions in which it will be used. When in doubt (and when your budget and weight restrictions permit it), use a larger gauge wire. Your resistance will decrease and your efficiency will increase.

Be careful when tightening bolts or nuts on the terminals of your battery cells. Too much force can easily strip the threads. Check with the manufacturer or vendor of the cells to see how much torque can be applied to the terminals. If this is a small battery, a torque wrench might be overkill. If this is a much larger battery, a torque wrench is a good idea to ensure you put enough force on the terminal connections without over-torqueing them. It should go without saying that if the battery is for a vehicle, such as a car or aircraft, this step should be crucial. A loss of power from a loose connection can have fatal consequences in such circumstances.

A torque wrench is a fairly precise tool. If you were assembling jet engines, I'd recommend using a pretty darn good one. For tightening connections on a battery's screw terminal, a cheap "eBay special" shipped directly from China is just fine for most projects. A $20 torque wrench might save you from stripping out the terminal on a $500 electric vehicle battery cell. Think about it.

The advantage of wire connections over busbars or spot welded connections is that you retain some flexibility and vibration resistance in your connections. If your battery will see any level of vibration, you should always use stranded wire. The more strands the better. Higher strand counts result in more flexible wires that withstand increased motion and vibration without failing.

When possible, it is often helpful to begin with parallel connections before performing series connections. This helps simplify the battery construction process and reduces the chance for careless mistakes that can have large consequences. If you can, avoid going back and forth between making parallel and series connections. Instead, try to connect all of your parallel groups first. Then when you have all of your parallel modules assembled and connected, move on to connecting those groups in series.

Depending on the design of your battery and the shape of its enclosure, it might not always be possible to construct all parallel groups first. "Parallel groups first" is not a hard and fast rule. It's only a suggestion that can often simplify the building process and help avoid mistakes.

Wire connector connections

Some batteries, mostly RC lipo batteries, will come prepackaged with charge/discharge wires and connectors already installed. This makes battery building much easier.

Instead of having to directly connect the individual cells of each battery, you simply need to use wire and matching connectors to join together individual packs. The number of batteries wired in series and parallel will of course depend on the requirements of your project. RC lipo batteries are widely available in battery packs up to approximately 30 V and 8 Ah. Higher voltage or capacity packs also exist, but aren't as common. An RC lipo brand known as Multistar produces higher capacity battery packs of up to 20 Ah.

For most smaller batteries under approximately 8 Ah, you won't have to make any parallel connections at all. A simple chain of series connections will be enough to build a battery for your system. For example, if you need a 63 V and 6 Ah battery, you could connect two 6s and one 5s RC lipo packs, each of 6 Ah, together in series. That would create a single 6 Ah battery with 17 series cells for a nominal voltage of 62.9 V.

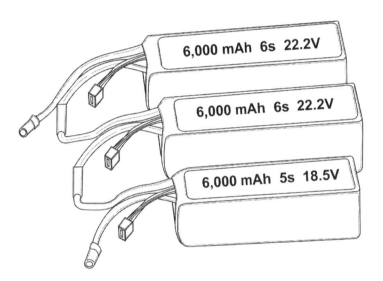

To create this battery, you'd only need to add wire connections between the discharge wires of each pack. They could either be plugged directly to each other, or you could fashion a new wiring harness to create cleaner looking wiring.

Remember of course that RC lipo batteries must be balance charged at least semi-regularly if you are not including a BMS. They also require voltage alarms or other voltage meter devices to ensure that they aren't discharged below a safe voltage. If you don't require the super high current available from RC lipo batteries though, you should consider installing a BMS to take advantage of the BMS's charging and discharging protection. The balance wires on RC lipo packs can be connected directly to a BMS, which we'll describe near the end of this chapter.

Spot welded connections

Depending on your cells, spot welding may be the only option available to you. This includes many cylindrical cells like 18650s. Spot welding has many advantages over other connections methods including its permanence and robust electrical connections. Good spot welds should never loosen up, unlike wires or busbars which can loosen over time under certain circumstances.

As previously discussed, there are two main types of spot welding electrodes: flexible handheld electrodes and rigid mounted electrodes. The first type will give you more freedom in the build process, while the second type can often result in more uniform welds as they tend to apply the same spring loaded pressure to each weld. Both work well, and the choice will often be decided by the size or shape of the battery you are building. Square packs lend themselves nicely to rigid mounted electrode arms, while odd shaped packs sometimes require flexible handheld electrodes.

The major difference in the process between using the two different types of spot welding electrodes is the order of operations. Depending on the shape of the battery, performing all parallel group welds first is sometimes more difficult or impossible when using spot welders with stubby electrode arms. This is because they limit the reach of the electrodes and the parallel groups may be too thick to allow the series connections to be made. In such a case, the battery must be built row by row, performing both series and parallel connections together. This is perfectly acceptable, but more care must be taken to not get confused and make an incorrect connection.

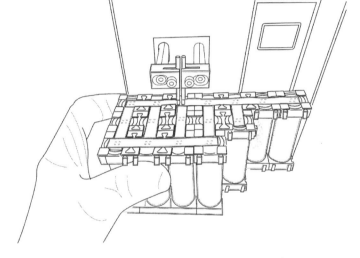

Another big advantage of handheld welding probes is that that they allow you to not only glue or otherwise arrange your battery entirely before you begin welding, but they also allow you to leave it on your work surface. The stubby arm electrode welders require you to physically lift your battery up to the welder and then raise the arms up by lifting the entire battery to engage the welder. The arms have a switch that fire the current pulses when the arms are lifted to the correct height.

If you're building a small battery, lifting it repeatedly can be manageable. When you're working with hundreds of cells in a single battery, it can become unwieldy. At a certain point, you end up manipulating a battery that is bigger and heavier than the welder itself, which would be almost comical if it wasn't so tiring and frustrating.

In some cheaper hobby-style welders, it can be helpful to add something heavy on top of the welder. The spring tension of the arms is often adjustable, but it can sometimes be close to the weight of the welder itself. That means the entire welder can occasionally lift up when you use the battery to raise the welding arms. A few additional pounds or kilograms on top of the welder helps to steady it.

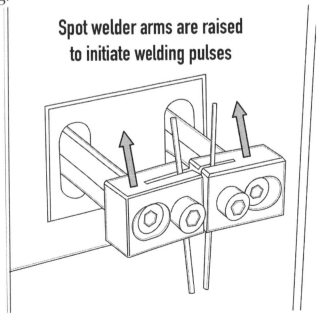

Spot welder arms are raised to initiate welding pulses

Most welders with stubby arm electrodes can work with either a foot pedal or by raising the electrode arms. Personally, I prefer to raise the electrode arms. I'm already lifting the battery up to the arms to begin with, so I like to use the same motion to control the timing of the welding pulses. Using the arms to engage the welding pulses also ensures an equal pressure at each weld. Other people prefer to use the foot pedal switch to engage the pulse. It's more of a comfort thing, and you should try out both options to see what feels best for you.

For hand-held welding probes, the foot pedal is usually the only option, though some probes come with a button mounted on the probes themselves. The button is often more convenient. We're not monkeys, we don't have the same motor skills with our feet as we do with our hands.

If you have many cells that are being welding in parallel groups in straight lines, a cell holding jig can be quite handy, especially if you aren't using the plastic cell holders. This is great way to hold your cells while you weld parallel groups, and then you can hot glue the parallel groups together to weld them in series.

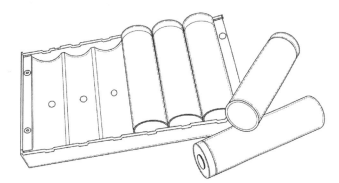

A cell holding jig usually has a series of half cylindrical cutouts and a space to lay a nickel strip on top of the cells. Mine came with magnets that help to hold the cells in place while I'm welding them. They are available online from sites like eBay and AliExpress, and many welders will come bundled with accessories including a jig.

If you can't find a jig, you can always make one yourself with a little ingenuity. Drill some appropriately sized holes in a piece of wood or plastic block, then cut it in half. You can even drill and countersink some little neodymium magnets in there for an even more effective jig.

When spot welding lithium cells, there is an optimum number of welds. Too few will result in increased resistance while too many can introduce unnecessary heat to the cells. For cylindrical cells like 18650s, I generally use 6-8 welding points, which is 3-4 welds when using two side-by-side welding probes. This has proven to be an acceptable number of welds.

Some devices such as computers and other electronics will often employ only a single set of welds in their battery packs. However, these devices generally use very low power. If your device is designed for low power, such as less than 1 A per cell, then a single weld may be enough. For devices that require higher power, an additional weld or two can give you two to three times the current carrying capacity. That's important for high power devices.

When you perform multiple welds on a single cell, its best to give the cell a few seconds to rest between each weld. A single weld creates very little heat on the cell. You can usually touch the point immediately after the weld and it will feel slightly warm. If you perform three or four welds in the span of a couple seconds, you'll notice the point becomes much hotter. This is because the heat builds up faster than it can dissipate into the air.

When possible, wait at least a couple seconds before repeating a weld on the same cell. This can be as easy as moving on to the next cell in the line before coming back and giving any cell "seconds" or "thirds", so to speak. That way each cell only gets one weld at a time and they all cool very quickly while you move on the next cell down the line.

We discussed the importance of designing your nickel strip layout so that there is enough current carrying capacity in every series connection. This is where that theory plays out in reality. Ideally you'll have multiple connections between each cell. If not, and you had to go with the stacking pyramid method, make sure that you're doing these stacks one at a time.

Start with the longest piece first, which will form the base of the pyramid. Perform all of your welds on that base piece, then add the next shorter piece. Weld that piece completely and continue adding progressively shorter pieces in this way until you have completed your welding pyramid.

One last little quick tip for spot welding with nickel strips. You already know you should be wearing gloves. Consider carefully the type of gloves you use. Fabric gloves like mechanics' gloves can often catch the sharp edge of a piece of nickel, especially when the nickel is cut with a pair of scissors that can curl the edge a bit. When you lay down a piece of nickel on your battery, it can be easy to catch the edge on your glove and accidentally drag it across your battery. This is a great way to create a short circuit. I prefer to use latex or nitrile gloves when I do battery work. They are less likely to catch on the nickel and they give me better dexterity. They can make your hands get sweaty though, so consider all your options.

Series vs parallel connections

Both series and parallel connections are important, but they don't necessarily need to be treated the same. In fact, they don't even need the same style of connections.

Series connections are where all the current flows, so they are in a way "more critical" than parallel connections. Both types of connections are important, but series connections must be designed to carry the highest amount current that the battery will see. Every series connection must be as strong (or preferably stronger) than the minimum requirement to carry the current that the pack will supply. Just like the weakest link in a chain determines the chain's strength, the lowest current carrying capacity series connection decides how much current can flow through the battery without overheating.

In circuits, Kirchhoff's current law states that the current at every point in an electrical loop is equal. That means the current at every series connection is equal. If one series connection is narrower, weaker, or uses less conductive material, it will have a higher resistance than the rest of the connections. That's the point that will heat up first in the battery and begin to limit its potential.

For example, imagine an 8s1p battery (shown in the diagram) made from cylindrical cells, which would be eight cells connected in series. If every spot weld was made with two layers of 8 mm wide 0.15 mm thick nickel, the series connections would be able to support around 10 A of current continuously. But what if one connection used only a single layer of nickel while the others all used two layers? That series connection would only be able to support around 5 A of current. If a 10 A load was connected to the battery, all the series connections would experience the same 10 A of current, but that single layer connection would heat up quickly due to its higher resistance.

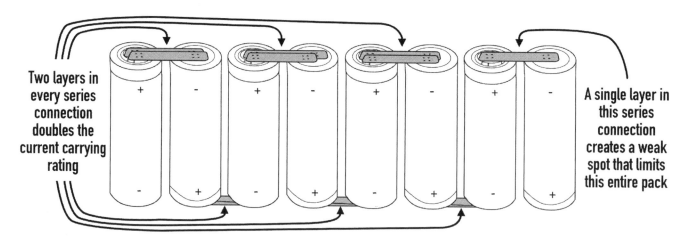

Two layers in every series connection doubles the current carrying rating

A single layer in this series connection creates a weak spot that limits this entire pack

That was a bit of an extreme example, where one series connection was half as strong as the others. As long as all of your series connections are *strong enough*, it doesn't matter if one has a bit higher resistance than the others. But if they all have four layers of nickel except for a single connection that you forgot and only gave one layer of nickel, your battery might as well have been built with one layer of nickel everywhere.

Parallel connections, on the other hand, will see much lower current. Likely hundreds of times lower current. Why? Because all of the current is flowing "downstream", so to speak, through the series connections. The only way current will flow "across stream", which would be between individual cells in the same parallel group, would be if one cell becomes slightly unbalanced.

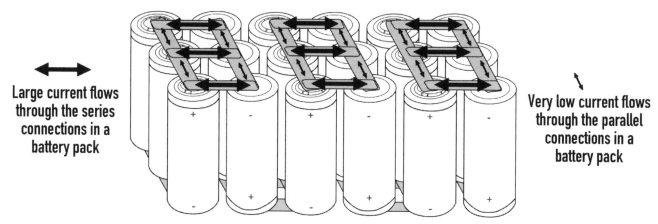

Large current flows through the series connections in a battery pack

Very low current flows through the parallel connections in a battery pack

During discharge, the cells in a parallel group will be working more or less equally hard. That is to say, they'll be providing the same current. But small differences in their internal resistances might mean that one cell provides a tiny bit more current than the others. During discharge, a small amount of current will therefore automatically flow from the stronger cells to the weaker cells inside of a single parallel group. This is likely on the order of milliamps.

Think of it like three really strong men and one medium strength man all carrying a refrigerator. They'll all be supporting their corner, but the strong guys might need to lift just a little more on the smaller guy's side to help him lift it the same amount.

So because there is only a very small amount of current flowing through parallel connections (i.e. between the cells in a parallel group), parallel connections can have very low current carrying connections. Cylindrical cells are often spot welded with the same nickel strip in parallel as they are in series. However, a single thin piece of nickel would be more than enough for parallel connections, assuming the cells in the parallel group are connected one-to-one in series with the cells in the next parallel group. All of the heavy current would flow through the strong series connections, and a thin wire soldered to the nickel strip would be enough to create parallel groups that can self balance each other.

There are two main reasons why cylindrical cells are often spot welded in parallel groups with the same heavier nickel as used for the series connections. When the parallel groups are created first, strong welded connections help hold the cells together. Secondly, stocking and swapping different connection materials adds more complexity to a large battery building operation. The cost difference is usually inconsequential for small batteries, meaning it's just easier to use the same material for all connections.

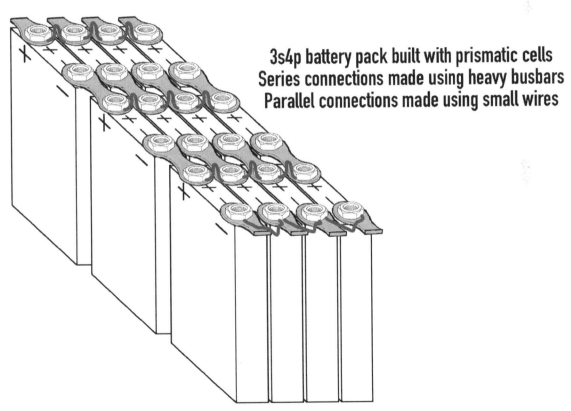

3s4p battery pack built with prismatic cells
Series connections made using heavy busbars
Parallel connections made using small wires

When you're working with bigger cells, and especially cells that are spaced farther away, the cost savings can begin to add up. For packs with many hundreds of kilowatt hours made using prismatic cells for electric vehicles, smaller parallel connections can save a lot of heavy duty wire. You can use the large gauge wire for making strong series connections and save the smaller gauge wire for parallel connections.

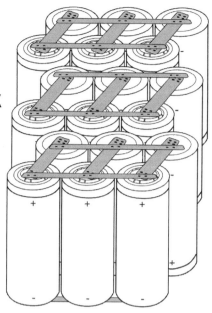

6s3p battery pack with thin parallel connections and thick series connections

To do the same with cylindrical spot welded cells, you'll want to weld up the series connections first, then go back and either spot weld or solder on smaller connections using thin nickel strip or wire. Remember to check that all of your cells are at equal voltages before connecting the cells in parallel, but you should be checking for that anyways regardless of the order of the connections.

Connecting the BMS

Most lithium batteries use some type of Battery Management System (BMS) to protect the battery during discharging and to balance the cells during charging. If you aren't using a BMS, you'll need to use a balance charger, which we'll cover in more detail soon in Chapter 13.

There are thousands of designs out there for commercially available BMS units. Many of them are fairly unique, which makes it hard to write one absolutely definitive guide for correctly wiring every single BMS in existence. However, most BMSs are similar and follow the same general wiring structure, so these instructions should guide you well for the vast majority of BMSs. If for some reason you encounter a BMS that has odd markings or doesn't seem to match what you've learned here, just contact the vendor or manufacturer. BMSs usually come with wiring diagrams for exactly this reason.

Your BMS will have a few thicker wires (or pads to solder on the wires yourself). These are the discharge and charge wires. They are thicker because these are the wires that will carry the largest discharge current and the somewhat smaller but still not inconsequential charge current. Your BMS will also have a number of thin wires. These are the balance wires. I like to start by first connecting the main charge and discharge wires.

Connecting charge and discharge wires

Most BMS units generally have B-, P- and C- wires (or pads to solder those wires). Occasionally there will be a B+ wire as well, but that is rarer. I say "generally" because, again, every BMS can be different and standards often aren't respected in the BMS industry. It is important to double check the wiring diagram for your specific BMS to confirm that everything in this chapter is true for your BMS.

The B- wire is generally for the negative terminal of the entire battery. That's also the negative terminal of your first parallel cell group. The B- wire should be wired directly to the negative terminal of that first cell group. It is important to connect the B- wire in such a way that all cells in the first parallel group can supply current equally. Connecting to just one end of the parallel group would force the parallel connections closer to the wire to carry more current. It is better to connect the wire to all cells, if possible.

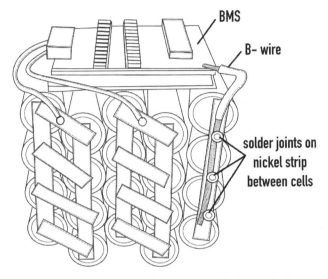

BMS

B- wire

solder joints on nickel strip between cells

This can be done by soldering it in between the nickel strip of every cell on spot welded batteries, or connecting it to each terminal on bolted connection prismatic cell batteries. The only exception to this would be if the battery discharge rate will be low enough that the parallel connections can sufficiently carry the current. Generally though that would be a very low power, long life battery.

The P- wire is generally the negative discharge wire for the pack. That means that it will be plugged into whatever device the battery is powering, such as your lights, motor speed controller, voltage converter, etc. This wire doesn't connect to your battery cells at all. It goes straight from the BMS to the negative terminal of whatever device you're powering.

The negative discharge wire on the BMS is almost always marked P-. However, I have seen the negative discharge wire on the BMS marked as PD- on a few BMS units. That is very rare though. It's almost always marked P- on most BMS units.

The C- wire is generally the negative charger wire. It won't connect to your battery at all, but will instead go straight from your BMS to the negative side of your charger's output. On the very few BMS units I've seen with a PD- wire for the negative discharge, the negative charger wire was marked as P-. That's confusing, I know, since P- is normally the negative discharge wire. What can I say? Go complain to the Chinese board designers and tell them to stick to the standard conventions.

If the BMS has a B+ wire, it is generally the positive discharge wire. It would be connected directly to the main positive terminal of the entire battery, which would be the positive terminal of your last parallel group in series. If your battery is a 13s battery, this would be the 13th parallel group.

When I connect wires to the BMS pads or lengthen the existing BMS wires by soldering on longer wires, I always like to put connectors on the ends of the wires first. This is so that there aren't exposed, bare ends of wires dangling around while I'm working on the battery. This would apply to your P- and C- wires. Your B- wire (and B+ wire if your BMS has it) will connect directly to the battery and won't need a connector. You can of course include a connector if you'd like to be able to disconnect the BMS from the battery, but that is purely optional.

One case where you might want a BMS that can be disconnected would be if you wanted to use the battery at a discharge rate that was higher than the BMS is rated to handle. Removing the BMS would allow you to discharge directly via the battery's main positive and negative terminals. There are two things to be careful of in this scenario though. First, make sure your battery cells can handle whatever high current you're trying to draw from them. And second, make sure that you use some type of meter or alarm to ensure you don't over discharge the cells, as the BMS won't be able to cut off the discharge when the cells are empty if it isn't connected.

After your main BMS charge and discharge wires are connected, you still have to connect additional positive charge and discharge wires directly to your battery as well as connect the balance wires. We'll start with the positive charge and discharge wires.

The BMS usually only connects to the pack using thick gauge wires on the negative terminal. Your main pack negative discharge wire will come from your BMS, but you still need to add the main pack positive discharge wire. Make sure to add a connector to the end of this wire as well before you connect it to your battery. It'd be a shame to create a short when you're this close to finishing your battery construction.

The main positive discharge wire should connect to each cell with sufficient current carrying capacity, just like the B- wire. Again, on a spot welded pack this can be achieved by soldering the main positive discharge wire to the nickel strip in between each cell on the last parallel group. On a bolted connection pack, the wire can be connected to each bolted terminal of the last parallel group.

The only time it would be acceptable to connect the main positive discharge wire to only one cell in a parallel group is if the discharge current will be low enough for the parallel connections to handle it. In the case of a parallel connection made from a single piece of nickel strip rated for 5 A, you could connect the negative and positive discharge wires to only one place on the first and last parallel group as long as won't draw more than 5 A from the battery.

The main positive charge wire will connect to the same place on the battery as the main positive discharge wire. The battery is charged and discharged via the positive terminal of the last parallel group. The positive charge and discharge wires could theoretically even be a single shared wire. But for practical purposes, a second wire is often used to avoid disconnecting the positive discharge wire from the device it powers. With separate wires, you could charge the battery while it was still connected to another device. This would obviously be useful for large batteries such as those powering a car or a home energy system. It would be annoying to disconnect those batteries each time to charge them.

The positive charge wire could also be spliced into the main positive discharge wire to avoid connecting directly to the cells, if this was desirable. If you are soldering your wires onto nickel strip, this method can help avoid adding unnecessary heat to the cells. You should of course try to solder on the nickel between cells and not directly above them in order to keep the point of heat further away from the cells. Even so, reducing the number of wires soldered to the nickel near the battery terminals is preferable.

Because the charge current is often much lower than the discharge current, it can sometimes be acceptable to connect the positive charge wire to only one cell in the last parallel group. For example, electric bicycle batteries often only charge at levels of around 2-4 A, but the nickel strip joining the parallel groups is often rated for 5 A. In this case, it would be perfectly fine to connect the positive charge wire in only one place on the parallel group. If we were charging the same battery at 10 A though, the charge wire should be connected in at least two places to avoid any segment of the nickel needing to carry more than 5 A.

Connecting the balance wires

That concludes the thick wires from your BMS. Now we must connect the balance wires on the BMS. The balance wires connect to each parallel group and are used by the BMS to monitor the voltages of each parallel group during discharging and to balance each parallel group during charging.

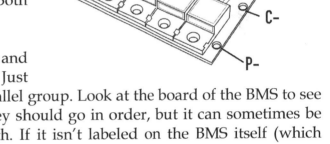

In nearly every BMS, there will either be the same number of balance wires as parallel groups or one more wire than the number of parallel groups. Both are common, so be prepared for either.

If there are an equal number of balance wires and parallel groups, the wiring is very simple. Just connect each wire to the positive end of each parallel group. Look at the board of the BMS to see which wire belongs to which parallel group. They should go in order, but it can sometimes be confusing to understand which end to start with. If it isn't labeled on the BMS itself (which would look like B1, B2, B3... near the wires), then check the BMS's wiring diagram.

Sometimes it won't be labeled, but all the balance wires will be white or black, with a single red balance wire on one end of the group of balance wires. Generally speaking, that red wire will be the highest numbered parallel group in the battery and each successive wire will belong to each lower parallel group number. Always double check this with the wiring diagram though.

If there is one more wire than the number of parallel groups in the battery, the same rules apply, except that the first wire will actually start at the negative terminal of the first parallel group. Then, the rest of the wires will continue as normal on the positive terminals of each successive parallel group until you reach the highest numbered parallel group. In a 5s BMS with six balance wires, the balance wires should be labeled as B1-, B1+, B2+, B3+, B4+, and B5+ with the sixth and final wire connecting to the positive terminal of the fifth cell group.

You must use the appropriate BMS for your battery based on the number of parallel groups in series. If you have a 10s battery, you cannot use a 13s BMS and just leave the last three balance wires disconnected. The BMS will be expecting to the see the voltage of those missing cells. When it doesn't detect them, it won't allow the battery to provide current.

You also can't use a BMS for one fewer parallel groups in series, even if it has an extra balance wire. For example, you can't use a 5s BMS with six balance wires on a 6s battery. It just won't work.

Your BMS will need to be securely mounted to your battery or the enclosure of your battery to keep it from moving and disconnecting any of the numerous wires. Smaller BMS units are typically a bare board, and should be insulated or covered if they will be mounted to the battery directly. An accidental short could occur if the BMS were to bridge the terminals of two adjacent parallel groups.

I often use kapton tape to mount smaller BMS units to batteries. Kapton tape is non static, non conductive and is quite sticky without being gummy like some electrical tapes.

For batteries that will experience vibration, such as for an electric skateboard or electric bicycle, I like to mount a thin piece of foam between the battery and the BMS board to reduce the shock loading on the BMS. Batteries cells can usually handle the vibration, but the more fragile printed circuit board of the BMS could be damaged under extreme impacts or shock loading.

Larger BMS units that are rated for higher power are often enclosed in their own aluminum case to protect the BMS and aid in heat dissipation. The case often has bolt holes for mounting the BMS. Make sure to keep the aluminum case away from the terminals of your battery to prevent a short circuit.

If you choose to skip the BMS altogether, your main charge and discharge wires will be much simpler. You obviously won't have any BMS to stand in the middle of the circuit, so you'll just connect the charge and discharge wires to your first and last parallel groups. The negative charge and discharge wires connect to the negative terminal of your first parallel group. The positive charge and discharge wires connect to the positive terminal of your last parallel group.

You don't actually need two sets of wires - the same pair of positive and negative wires can be used for charging and discharging. But depending on the type of connectors you want to use and whether you want to leave your battery connected to the device it is powering, it can be useful to have a separate charging and discharging port.

Adding a balance connector

If you don't include a BMS in your battery, then you must include a balance connector. The balance connector is connected just like the balance connector on the BMS. A single thin wire is connected to each parallel group of the battery. This allows the battery to be balanced during charging, which we'll discuss in more detail in Chapter 13.

Even if you do choose to use a BMS in your battery, adding an extra balance connector might be a good idea. By having the balance connector external to the battery, you can monitor the voltages of the parallel groups without having to open up your battery. This can be helpful down the road when you want to ensure that the battery is still healthy after many cycles. It can also be good for confirming that your BMS is still working. If your BMS ever dies or the balancing function were to fail, you might not realize it until your battery becomes very unbalanced and starts having issues. An external balance connector can help you verify that the BMS is still working fine.

BMS wiring guide for a 10s BMS with 10 balance wires

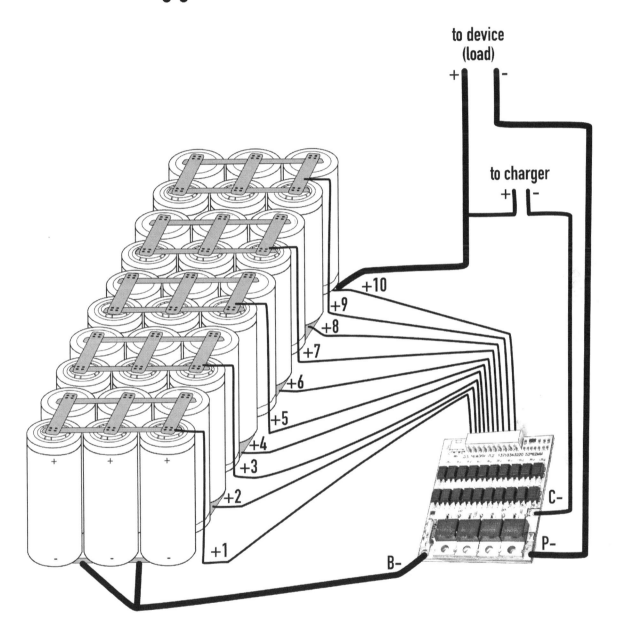

Performing final quality and performance checks

At this point your BMS is connected and your battery is finished! Well, electrically speaking anyways. We've still got to seal it up and make it pretty, but that's for the next chapter. Now it's time to double check that everything has been connected correctly.

Use a digital multimeter or a voltmeter to check the voltage of your battery through both the charge and discharge wires. You should get something reasonable in the range of your finished battery. A 36 V battery might not read exactly 36 V if the cells weren't fully charged to begin with. Many li-ion cells ship from the factory at 3.3 V per cell, so a 10s 36 V battery built from these cells should read 33 V at both the charge and discharge connectors. If for some reason you get an odd number, especially a number that is too low, it's probably due to an incorrect connection somewhere with the BMS. Double check all of your main connections and balance connections.

If you get a reading of 0 V or an open circuit reading, it may be a BMS issue, but it may also be a cell connection issue. To determine which, measure the voltage from the negative terminal of your first cell group to the positive terminal of your last cell group. If it's still 0 V or an open circuit, that means you likely forgot to make a series connection somewhere. If you get the appropriate voltage then you have a problem with your BMS, likely an incorrect BMS connection somewhere.

If you've checked everything and still get an incorrect voltage reading that isn't within the range of your fully charged to discharged voltage, you might just have a bad BMS unit. This happens occasionally, especially with cheaper "China specials". If you're going to buy a cheap BMS, you might want to buy two just to have a replacement. But if you're going to buy two, you might as well spend twice as much on a good one to start with.

Checking the voltage is a good way to make sure everything is correctly connected, but it won't tell you how well everything is connected. For that, it's best to test your battery with a real load. If possible, testing with the actual load it was designed for would be optimal. For some applications though, it's not possible to put the bare battery in the device, or the device isn't ready yet.

If your device is available and you can test the battery with it, try running the battery under load for a few minutes to make sure you don't feel connections on the battery heating up. A little bit of heat is fine. Under different loads, many cells will warm up to the 35-50 °C range. Unless you're building a pack that is designed to operate under very high power, you don't want it to get too much hotter than that.

An infrared or laser thermometer is great for checking the temperature of different parts of your battery. They can be bought for less than $5 on sites like AliExpress. Pay careful attention to your series connections to make sure they aren't getting too hot either. A bright red glowing piece of nickel strip would be a dead giveaway that you've messed up, but even just a single series connection that is much hotter than the others can be an indication that that connection isn't sufficiently strong (i.e., it needs more current carrying capacity).

If you can't use your intended device, there are other ways to test a battery. 12 V batteries can be plugged into a wide array of 12 V DC devices, such as lights or heaters. Other voltages don't lend themselves as easily to powering the devices you have laying around, but they can be plugged into a voltage converter to output 12 V DC for a similar effect.

You can also create a "dummy load" by using a heating coil, power resistors or a chain of light bulbs to create a resistive load for the battery. Halogen light bulbs and power resistors are often used to build homemade battery discharging devices. A quick google search for either will find many different sets of instructions to build a battery discharger. I even posted a video on the EbikeSchool.com YouTube channel showing how to make an adjustable voltage halogen lightbulb discharger.

Testing your battery under a load isn't crucial, though it's a nice extra step to ensure quality. But as long as your battery is providing the proper voltage from its charge and discharge connectors, then you're ready to move on to the next step.

Chapter 12: Sealing the battery

Once you're finished with your connections and confirmed that your battery is in good working order, you can go ahead and seal it up. In my opinion, all lithium batteries should be sealed in some way. An unsealed lithium battery with exposed terminals is simply an accidental short circuit waiting to happen.

The only case where I commonly see lithium batteries left unsealed is in the DIY powerwall community. They generally use snap together plastic cell holders which create a nice battery case and keep the cell terminals lifted off of the surface they rest on, but they don't actually cover the exposed terminals.

I'm not sure why these folks don't seal their batteries. Maybe they want ease of access to replace bad cells, since they usually build their batteries with salvaged cells. Maybe they want the benefits of passive air cooling and don't want to design in an active airflow system. Maybe they just figure that the batteries are up on a shelf or hidden in a closet and therefore they're probably safe.

Either way, such batteries with exposed terminals represent a fire hazard if anything were to ever short them. A mouse in the closet would turn pretty crispy if it stepped in the wrong place and shorted some high capacity home energy storage batteries.

The method of sealing your battery will depend on a number of factors including the type of cells you used, the size and shape of your battery, and the environment in which your battery will be used.

Hard cases

Prismatic cells are usually very easy to seal. Their only exposed terminals are small threaded rods or nuts and often come with special rubber covers to insulate them. If they were connected with insulated wires, then the wires don't need to be covered. If busbars were used to connect prismatic cells, the exposed busbars should be covered somehow. Often a rigid case is built around the prismatic cells to cover the cells and busbars. A hinged door allows easy access while still keeping everything protected and safe.

Pouch cells are also good candidates for hard cases. The delicate cells can be easily punctured. Rigid cases such as Pelican cases provide protection against accidental damage. Pelican cases are fairly expensive though, and cheaper plastic cases or hard plastic lunch boxes also work well as battery cases.

Even though hard cases are recommended for pouch cells, due to the lightweight and energy dense nature of these cells, they are often used in instances where weight and volume are minimized. In cases of drones and light aircraft, rigid cases add unwanted bulk. When using pouch cells without a hard case, extra care must be taken to protect the cells from accidental damage.

Heat shrink

Heat shrink is a good method for sealing many different types of batteries. It is often used as a first step in the sealing process, then followed up by a hard case to add extra protection.

Batteries made from cylindrical cells require careful sealing to keep the battery safe from short circuits. The ends of the cells are nearly covered in nickel strip connections, making short circuits much easier, depending on the battery layout and orientation.

Cylindrical cell batteries lend themselves well to heat shrink wrapping, depending on the shape of the battery. Large diameter heat shrink wrapping is commonly available online in sizes large enough to heat shrink wrap a child. Not that you should do that.

For batteries that will be used in a high vibration environment like electric skateboard and bicycle batteries, it's a good idea to wrap a layer of foam sheet around the battery first before heat shrink wrapping it. This will give some extra padding and protection from bumps and shock loading. Arts and crafts foam sheets between 1-3 mm thick work well for this purpose and are quite inexpensive. Some people use cheap yoga mats, though that foam is likely thicker than necessary and will trap more heat than thinner foam. Depending on your application though, thicker foam might be an advantage.

Thin fiberglass boards are also used for surrounding a battery and providing rigid protection. Fiberglass doesn't absorb impacts as well as foam, but it helps to spread out and sharp bumps over a larger area of the battery.

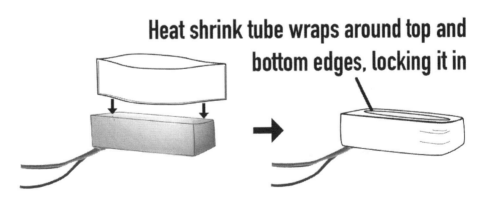

Heat shrink tube wraps around top and bottom edges, locking it in

Heat shrink wrapping is easiest for square shapes where it can wrap around two opposing edges, effectively locking itself in place. For triangular batteries or any shape with steep inclines, heat shrink can be a bit more difficult to use. The problem is that a single loop of heat shrink wrap often can't shrink small enough to grip the opposing edges, and it will therefore slide down the incline. High-ratio heat shrink is available that can help solve this problem, but it's harder to find in such large diameters.

Heat shrink on inclined surfaces can loosen and slide off towards the narrower side

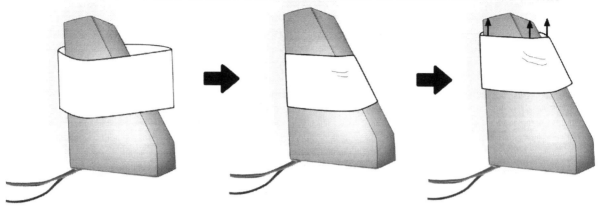

One trick for heat shrink wrapping triangle shapes is to use progressively larger sizes of heat shrink as you go further along the incline of the battery. The larger sheets will shrink down on the smaller sheets that are already in place and help to lock them down.

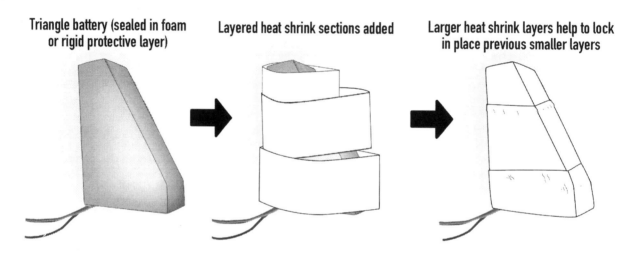

Triangle battery (sealed in foam or rigid protective layer)   Layered heat shrink sections added   Larger heat shrink layers help to lock in place previous smaller layers

Using heat shrink wrapping loops that are oriented 90° to each other can also help lock the heat shrink down on shapes with inclined planes.

Heat shrink layers emplaced at 90° angles to each other

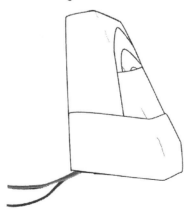

Many people find it difficult to determine the correct size of heat shrink to use for their battery. Large diameter heat shrink is measured by the half-circumference instead of by the diameter. It's also usually listed in metric sizes. So a piece of 150 mm heat shrink isn't a circle that is 150 mm (6 inches) in diameter. Rather, when laid flat, it measures 150 mm across. That flat width is equal to half of the circumference of the heat shrink when it is opened into a circle.

Calculating the size is fairly easy once you know those points. All you need to do is measure your battery to find half of its perimeter. If the battery is a regular square or rectangle shape, measure the top and one side, then add those two numbers together. That will give you half of the perimeter of your battery in the direction you measured. Because heat shrink generally shrinks 2:1, the largest piece you could use would be twice that number you just calculated, since it would shrink to exactly the perimeter of your battery. Obviously we don't want to cut it that close – we'd rather the heat shrink try to shrink smaller than our battery and thus squeeze our battery tight. So our optimal size would be between a bit under twice our half perimeter and a bit over our half perimeter (so it can still slide over the battery).

Maximum heat shrink size = 2 × (length + width of battery) – "a little bit"

Minimum heat shrink size = length + width of battery + "a little bit"

Basically all these two equations are saying is that our heat shrink size should be something between our perimeter and half of our perimeter.

Let's do an example. Let's say our battery is the rectangular one from the diagram two pages back. Let's say it measures 70 mm x 75 mm x 150 mm. To calculate the size of heat shrink we'd need to cover it around the long dimension, as shown in the diagram, we'd add the length and width giving us:

Minimum heat shrink size = 75 mm + 150 mm + "a little bit" = 225 mm + "a little bit"

So we know that our heat shrink needs to be a bit larger than 225 mm to fit around the battery. To calculate the maximum size, we will simply multiply that number by 2, which gives us 450 mm. So our heat shrink can be anywhere from 225 mm to 450 mm. Preferably we'd like something on the lower end of that scale to make sure the fit is as tight as possible. A 250 mm – 300 mm size would be great. And that's all there is to calculating large diameter heat shrink sizes.

When heating the heat shrink wrap, be careful not to apply too much heat. Many heat guns are designed for heavy duty purposes such as removing paint. These high heat settings can melt your heat shrink in less than a second. Start with a low setting and slowly increase the heat until you find the appropriate level. High powered hair dryers can also be used for heat shrink. I used my wife's 2,000 W hair dryer for years until I got a decent heat gun.

Cooling issues

Depending on your application, your lithium battery might not need any form of cooling. Most commercial lithium batteries in the few kWh range have no form of cooling. The lithium batteries used for devices such as drones, power tools, electric skateboards and electric bicycles are sealed in either heat shrink or hard cases and are used within power levels that don't require active cooling. Passive cooling that occurs when the case of the battery dissipates built-up heat to the surround ambient air is all that is necessary in most cases.

Some batteries have BMSs that cut their power when a certain temperature is reached. This is common in power tool batteries that are often used (or abused) at high power levels.

Cooling becomes more important at high power and when human lives are at stake. This is most commonly found in the automotive and aeronautical industries. Electric vehicles usually have active cooling systems using air, water, oil or other fluids to draw heat out of the battery.

The vast majority of the people reading this book will never need to consider adding active cooling to their battery. As long as the battery isn't abused and it is used in an environment with at least some form of passive cooling where air can pass over the battery case, it will likely be just fine. Thermal runaway in lithium cells becomes an issue at around 150 °C, yet lithium batteries in normal use shouldn't exceed 60 °C.

Generally speaking, as long as the case is exposed to air somehow, the batteries can sufficiently cool on their own. Power tool batteries might need to sit on the shelf for a few minutes when they hit their thermal cutoff limit. Electric bicycle and skateboard batteries usually cool passively just from the air that rushes around them while riding. Drone batteries are generally showered in the downwash of the propellers, which provides a form of passive/active cooling.

If you do require actual active cooling in your battery for high power applications, air is usually the best method. A sealed battery can have a cooling fan such as a computer case fan plumbed into the case with an exhaust port on the other side of the battery. Care should be taken to ensure that foreign objects don't enter the battery, but this method will provide sufficient cooling for most high power scenarios.

If you need water or oil cooling, your needs are likely outside of the scope of what one book can handle. You need a team of engineers designing your battery because lives are probably on the line. Applications like electric cars and aircraft have very sophisticated cooling systems. Building batteries for such applications should not be taken lightly.

Remember, adding more cells in parallel to create a larger capacity battery will help reduce the relative power required by the battery. Higher capacity batteries result in lower C rate discharging. This in turn results in lower heat generation. If your battery is approaching a thermal limit, there's a good chance that it simply wasn't designed with enough capacity or it wasn't designed with cells rated for a sufficiently high C rate.

Chapter 13: Charging lithium batteries

Charging lithium batteries isn't difficult. However, if charging is not done correctly then it can become dangerous. Many lithium battery house fires you hear about on the news occur during the charging phase. This doesn't mean you should be worried. It just means you should pay attention to the proper charging methods and employ them when appropriate for your battery.

The type of charger used for lithium batteries will depend on whether the battery has a BMS or not. However, the mechanism of charging lithium battery cells is the same, regardless of the battery construction.

Constant Current, Constant Voltage charging

Lithium battery cells charge by using what is known as a constant current, constant voltage (CC-CV) charging scheme.

A CC-CV charging scheme means that the first part of the charging period is a constant current phase while the second part of the charging period is a constant voltage phase.

During the constant current phase, electricity is supplied to a lithium battery cell with a constant current. This means the current is unchanging, even though the voltage will change during this period. A discharged lithium-ion cell might be at around 2.7 V when connected to a charger. A CC-CV charger rated for 1 A will supply 1 A of current to the battery cell, starting at a voltage of 2.7 V to match the voltage of the cell.

When the lithium cell experiences the current flow from the charger, the voltage will instantly increase, likely by around 5% or so. This is essentially the opposite of voltage sag during discharge. When the lithium cell remains connected to the charger, the voltage of the cell will slowly rise as the charge state of the cell increases.

This is where the CC-CV magic happens. When the voltage rises up to a set preset limit in the charger (usually 4.2 V for a li-ion cell), the charger will automatically switch from the constant current phase to the constant voltage phase. In the constant voltage phase, the voltage will remain constant at the fully charged cell voltage (4.2 V in this example) but the current will slowly drop. The current will continue to decrease as the last bit of capacity is offloaded into the cell. The charger will cut the current when it reaches a certain minimum amount, often around 100 to 200 mA. At this point, the lithium cell is fully charged.

The same process occurs in multiple cells when they are connected in parallel and series. Because parallel cells automatically balance each other, cells connected in parallel will charge together at the same rate. And just like how parallel groups connected in series will discharge at approximately the same rate when connected to a load, the same parallel groups connected in series will also charge at approximately the same rate when connected to a power source, which is the opposite of a load. That is to say, as the battery pack charges, all parallel groups will charge approximately equally.

The correct charging voltage is important to ensure that cells charge completely but do not overcharge. As we learned in Chapter 3, most LiFePO$_4$ cells charge to around 3.65 V while most li-ion cells charge to around 4.2 V, though a few specialty cells should be charged to 4.3 V or 4.4 V. To find the proper voltage of a charger for a battery built of many cells connected in series, you simply multiply the number of cells in series by the full charge voltage of the cells.

A 14s battery with li-ion cells will need a charger with a voltage of 58.8 V.

Total charge voltage = 14 cells in series × 4.2 V = 58.8 V

A 10s battery with LiFePO$_4$ cells will need a charger with a voltage of 36.5 V.

Total charge voltage = 10 cells in series × 3.65 V = 36.5 V

The number of cells in parallel will not affect the necessary charge voltage. A 14s li-ion battery will require a charge voltage of 58.8 V regardless of whether it has a capacity of 10 Ah or 50 Ah. What the number of parallel cells *will* affect is the current that the battery can be charged at. More parallel cells mean the battery can handle a higher charging current, just like it also means the battery can handle a higher discharging current. Of course with enough cells in parallel, you might reach the charging current limit of a BMS before the charging current limit of the battery cells themselves.

Charging with a BMS

Using a BMS makes the charging process much simpler. To charge a battery with a BMS, you simply plug the charger into the battery's charger port.

That's it.

Done.

The BMS takes care of everything. It starts by allowing the battery to charge from the charger using the bulk charging method. You'll remember from Chapter 9 that bulk charging is where the entire battery is charged as one unit by flowing current through all of the parallel groups of cells connected in series. Each parallel group sees an equal amount of current. Once the bulk charging process brings the first few cells up to full charge, the BMS then begins balancing the battery by draining energy from the parallel groups that reach full charge first (if using the top balancing method, which is most common). The BMS stops draining when all the cells reach full capacity.

But that's all taken care of automatically. From the perspective of the user, the charger is simply plugged into the battery and that's it. Nothing else to do but wait until the charging is complete.

There are a few types of chargers available for batteries with a BMS. They all work fairly similarly. The main difference is quality.

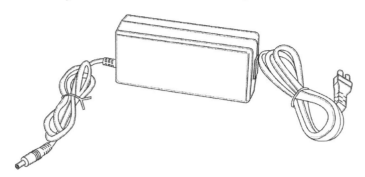

The bottom of the list begins with plastic case chargers. These are very cheap, usually in the $20-$40 range. They usually have no cooling fan, which limits them to fairly low current. Usually in the range of just a couple amps.

Plastic case chargers are produced to be affordable. There are some good ones out there, but there are also many that cut corners to save on costs. This is usually to provide a charger included with a battery or electronic device without having to spend too much on producing the charger itself.

Plastic case chargers can work fine, but the cheapest ones should be watched to ensure they aren't overheating. Unfortunately, this often happens quickly and without warning when something shorts inside the charger. Not only can this destroy your battery, but it can cause either the battery or charger - or both - to catch on fire.

If that sounds bad, then you might like the next class of chargers. These are the aluminum case, cooling fan chargers. As you might have already guessed, these chargers have an aluminum case and include cooling fans. They are generally more expensive, but can also handle higher current and power than the cheap plastic cased chargers.

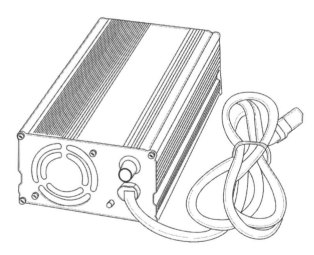

The higher cost and ability to handle higher current generally equates to better build quality, though of course this isn't always true. Generally speaking though, chargers with an aluminum shell and a cooling fan are both safer and higher quality chargers.

The next step up would be an adjustable, aluminum case charger with a cooling fan. These chargers are usually based on the same structure as the chargers I previously described, but include the option to adjust the charging voltage, current or both. The advantage of these is that you can choose a lower current for healthier charging or a higher current for faster charging when you're in hurry. You can also adjust the voltage to a lower level to help increase the lifespan of the battery. We'll talk about how that works in the next chapter.

At this level there are multiple versions of good quality chargers. One I particularly like that can handle anything from 12-72 V is the Cycle Satiator made by Grin Technologies in Canada. This charger is not only adjustable, but it's also programmable with a digital screen to allow you to add multiple charging profiles for different batteries. It's even waterproof. It actually doesn't have a cooling fan, but in this case that isn't a sign of low quality. It was simply designed to be incredibly efficient, thus not needing a cooling fan. It does get a bit hot when used at its highest power levels though, so be aware of that and make sure to give the charger room to breathe.

The charger you choose will ultimately be decided by your specific requirements. Low power batteries are often fine with cheap plastic chargers. Essentially every laptop and cell phone uses a cheap plastic charger to charge its lithium battery, so you shouldn't be afraid of them. The bigger quality issue comes into play when power levels increase, meaning higher voltages and higher currents. For higher power batteries, a better charger is a form of insurance against damage. For an extra $20 or so, a charger with an aluminum case and cooling fan is likely a good investment.

Charging without a BMS

Charging without a BMS is where things get a bit more complicated. Batteries without a BMS can still be bulk charged using the same chargers listed above that are meant for batteries with BMSs. However, only using a bulk charger over many charging and discharging cycles can result in the parallel cell groups in a battery becoming unbalanced from one to another.

Good quality lithium cells can handle many more charge and discharge cycles before becoming unbalanced, but eventually all parallel groups will unbalance if given enough cycles without a BMS to correct for this problem. One method used to avoid balance charging for as long as possible is to set the charge voltage slightly lower than the maximum rated charge voltage. For li-ion cells, this might mean charging to around 4.17 V per cell instead of 4.2 V per cell. By slightly undercharging the entire pack, the cells that eventually begin to overcharge will take longer to reach a voltage above their maximum rated voltage. Eventually though, only bulk charging with no balance charging will cause some cells to charge higher than their maximum rated voltage.

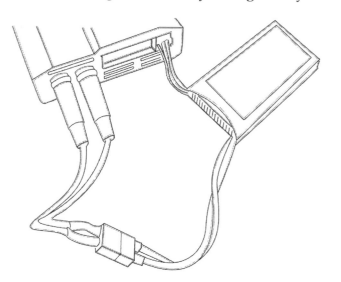

In order to avoid this case, a balance charger is used to charge the battery while also balancing the cell groups. Batteries without a BMS must have a balance connector wired to each cell of the battery. A balance charger requires the battery to be connected by both the balance wires on the battery and the charging wires (or the discharging wires, as they connect to the same location on the battery when there is no BMS involved).

The charger starts by bulk charging the battery via the charge wires (or discharge wires) until the battery approaches full capacity. At that point the balance charger works quite similarly to a BMS. The balance charger monitors the voltage of the individual parallel groups and bleeds off some capacity from the cells that reach full capacity first while continuing to charge the rest of the cells.

Remember, batteries without a BMS don't always have to be balanced charged. As I mentioned, good quality cells can last for many cycles while staying closely balanced. Higher current charges and discharges will cause cells to more quickly lose their balance. Low quality cells will also lose their balance quicker. The higher the current or lower the quality of the cells used in a battery without a BMS, the more often it should be balanced charged.

The best method, should you choose to go without a BMS, is to test your own battery to see how quickly it becomes unbalanced. You can always use a digital multimeter or voltmeter to measure the voltage of each parallel group via the balance connector on your battery.

Charging temperature

It is very important to respect the manufacturer's ratings for charging temperature. Charging lithium cells above or below their rated charging temperature can damage the cells and be potentially dangerous.

Charging lithium battery cells at low temperatures, below 0ºC, will damage or destroy the cell. At these low temperatures, lithium metal will permanently collect on the anode, reducing the capacity of the cell. Not only does this rob the lithium cell of capacity, but it can also increase the risk of the battery cell catching on fire if it is suffers an impact such as a drop or fall. Freezing temperatures aren't normally reached for most li-ion powered projects, but batteries that are left outdoors overnight or even during the day in very cold climates might experience this situation.

Electric vehicles often include safety mechanisms to prevent charging at very low temperatures. Some electric vehicles even employ built in heaters to warm the battery to a sufficient temperature before charging can safely occur.

For smaller batteries, bringing cold batteries inside and allowing them to warm up to at least 10ºC for a few hours will help avoid this damage during charging. It is always better to use a lower charge current when charging colder batteries.

Charging lithium battery cells at high temperatures can both reduce their capacity and potentially lead to thermal runaway. Charging a lithium battery cell produces heat due to the internal resistance of the cell. If the cell is already approaching a dangerous heat level, additional heat from charging can push it over the edge and into the realm of thermal runaway.

Remember that lithium batteries will have a wider safe temperature range for discharging than for charging. Always stick to the manufacturer's rated limits for charging temperatures.

Chapter 14: Increasing cycle life

Lithium batteries are expensive. Fortunately for us they normally last a long time, but anything that can be done to increase their lifespan helps delay a costly replacement. There are a number of useful tricks that can keep your battery cells healthier for longer and increase the number of cycles that your battery can provide.

Charge to a lower maximum voltage

Most lithium battery cells don't like spending a long period of time at higher voltages. In fact, it is recommended to store lithium battery cells at very low states of charge. The one major exception to this is $LiFePO_4$, which can tolerate staying at full charge voltage for longer periods of time with less cell degradation.

For most li-ion chemistries, keeping a cell at a high voltage causes damaging chemical reactions in the electrolyte of the cell. Over time, these chemical reactions slowly decrease the capacity of the cell. This can result in the premature death of the cell.

This doesn't mean that you won't get the number of charge cycles stated by the manufacturer of the cell though. What it does mean is that there's the potential to actually get more cycles than were originally stated, if you play your cards right.

Tests have shown that by charging li-ion cells to 4.1 V instead of 4.2 V, the cells could perform many more charge and discharge cycles. In some cases, cell charged no higher than 4.1 V perform over twice as many cycles.

In order to charge a battery to a lower voltage, you'll either need to modify your charger, or use a special charger that is adjustable. Even non adjustable chargers often have potentiometers inside for fine tuning the voltage. You shouldn't mess around with these if you don't know what you're doing. However, if you can get the info from the vendor on where to make the adjustment, you can turn down the voltage a bit. Keep in mind that this could void your charger warranty.

Adjustable chargers make this easier as they are designed to allow this very function. The only problem with this undercharging method is that a BMS that balances the cells at their full capacity likely won't ever enter its balancing mode, as the cells never reach their full voltage. For this reason, it's good practice to occasionally charge a battery to full voltage to allow the BMS to balance the cells. You might be able to find some BMSs that are designed for lower voltage balancing, but they are fairly rare.

The major disadvantage of this undercharging method is that you won't get the full capacity from your cells. For example, charging a li-ion cell to 4.1 V instead of 4.2 V means you'll only get about 90% of its full charge capacity. If doubling the life of your battery is worth cutting 10%

capacity off of each discharge though, then this may still be a good option. You can also play with the amount of undercharging you perform to find the right balance for your needs.

Interestingly, the opposite of this method also applies. By charging li-ion cells past 4.2 V, such as to 4.3 V, you can get more capacity than the cells are rated for. A 3 Ah cell might deliver 3.2 Ah. However, this not only drastically reduces the lifespan of the cell, but it can also be very dangerous. You should never charge cells higher than their rated voltage.

Discharge to a higher minimum voltage

Similarly, you can also use a higher low voltage cutoff. Lithium batteries have longer lifespans when they aren't discharged to low voltage levels under heavy loads. Li-ion cells are usually rated for discharging to 2.5 V, but stopping at a higher voltage will result in achieving more cycles from the cells.

This creates a similar problem to undercharging though, in that you don't get the full rated capacity of the cells for each charge cycle.

Many electric vehicle manufacturers use this method. Ever wonder how they can give a 10-year warranty on their batteries when most cells are only designed to last 1,000 cycles at most? They are often using only the middle part of the voltage range for their cells. In this case, they charge to a level slightly under the maximum rated voltage and stop discharging just above the minimum rated voltage. Not all electric vehicle manufacturers do this, but some have used it quite successfully.

Charge and discharge at lower currents

The lower the charge and discharge current of a lithium battery cell, the healthier it will remain. Even cells that are rated for very high discharge currents are still happier at lower current levels.

Fast charging is convenient but lowers the number of useful cycles you'll get out of your battery. When time isn't an issue, you should charge at lower rates to increase your battery life.

It can be difficult to discharge at lower rates if a specific application or device requires a fixed current. In such cases, increasing the capacity of the battery will result in relatively lower discharge rates. For example, a 2 A load on a 2 Ah battery cell is a 1 C discharge rate. But if you can use a 4 Ah battery instead, that same 2 A load results in just a 0.5 C discharge rate. The 4 Ah battery will last last for more complete charge/discharge cycles in this example.

Keep cells cool

We've discussed this multiple times so I won't spend too long on it. As we already know, heat is the enemy of lithium battery cells. Tests have shown that the higher the temperature at which lithium battery cells are stored and used, the fewer cycles they provide.

The sweet spot seems to be right around room temperature at 25 °C. Batteries will naturally heat up during charging and discharging, but you should do your best to store them at room temperature when not in use. Doing so will help prolong their useful lives.

While lithium batteries can generally provide good capacity at higher temperatures compared to other types of rechargeable batteries, the state of charge can have a big effect on their reduction in capacity. Performing complete charge and discharge cycles at elevated temperatures can quickly destroy the cell in just a few charge cycles. Conducting only 50% discharge before recharging can greatly improve the number of cycles the battery can perform at high temperatures.

Chapter 15: Disposing of old lithium batteries

When lithium batteries eventually wear out and end their useful lifetimes, they generally shouldn't be thrown away with normal household garbage. Laws regarding disposal of lithium batteries vary all over the world and even from state to state in the US. In many places, lithium batteries are considered hazardous waste.

In some locations it is permissible to dispose of small quantities of lithium batteries in normal municipal waste. In others, any amount of lithium batteries must be disposed of separately or sent to recycling centers. Still others have no laws at all regarding disposing of lithium batteries.

In some areas, lithium batteries are collected for disposal by incineration under controlled conditions. This prevents the potentially dangerous case of landfills filling up with lithium cells and starting massive fires, but is an unfortunate waste of such a valuable resource.

Even if your area permits disposal of lithium batteries, recycling is a much more sustainable option. Lithium is a limited resource. Its mining is both dirty and expensive. By recycling old lithium batteries, you're helping to save this resource and prevent further environmental damage caused by its raw collection.

Most areas have free recycling drop-off locations. These are often located at schools, universities, malls, post offices and other public places. It costs you nothing but benefits us all. If you don't know where you can recycle lithium batteries in your area, contact your local municipal waste company.

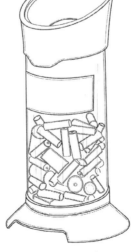

Whether you are sending the batteries off for recycling or just throwing them away in your household garbage, always prepare the cells first. Lithium battery cells should be discharged completely down to or past their minimum rated voltage. The cell terminals should also be wrapped in plastic or covered with tape to prevent a short circuit.

Chapter 16: Example battery building projects

There are an endless number of projects out there that can make use of DIY lithium batteries. To help reinforce some of the concepts we've learned in this book and to get you started down the path on your own project, I'm going to go over some example batteries for common projects.

One note in advance: these are going to be very broad overviews. They aren't going to get into the nitty-gritty details that are unique to building each project. Each project could be a book in itself. I'm just going to focus on the battery building and keep the overview at a fairly broad level. If you want to take a deep dive into these types of projects, then you should spend more time researching the specifics that go into the entire project and parts selection.

Ok, now let's get started.

5V backup battery and USB device charger

There are thousands of different backup batteries out there. Some are even available for cheaper than it'd cost to build one yourself. But just for fun, let's look at how we could make one ourselves.

First I'll choose a cell type. I could make a very small charger with a single 18650 cell. That's the cell format used in the common "lipstick style" cell phone battery chargers. But let's make something with a bit more capacity. I'll choose a li-ion pouch cell. A quick search turns up a 4 Ah pouch cell on AliExpress for $7. It even comes with wires already connected to the tabs, perfect.

My battery cell is 3.7 V nominal, but I need to output a constant 5 V to a USB port to charge USB devices such as a phone or e-reader. That means I need a small 5 V DC-DC boost converter. Another quick search on AliExpress and I've found a small DC-DC boost converter rated for 1 A discharge. It even comes with the USB port already attached. It has a low voltage cutoff (LVC) of 2.5 V which is also conveniently the minimum allowable voltage for my li-ion cell. I'll probably want to test that to make sure the LVC works, but it sounds good so far!

Lastly, I need a charging board. Back to AliExpress, where I've found a TP4056 charging module for $0.37. Shipping is another $0.35 though. Ouch, that's where they get you.

The TP4056 module has a micro-USB port to accept a 5 V input and will charge my li-ion cell up to 4.2 V. Exactly what I need!

Because I'm using a pouch cell, I need some type of case to protect it. One last search for "plastic electronics enclosure" brings up a pile of options. I'll choose one that fits my cell and has a little more room for the two electronics boards.

Now I've got all of my parts and I just need to combine everything. I'll start by soldering the wires from my pouch cell onto my TP4056 charging board. The pouch cell's red wire from the positive terminal gets soldered to the B+ pad on the charging board and the black wire from the negative terminal on the pouch cell gets soldered to the B- pad on the charging board.

Now I should test the charging circuit. I'll plug in a micro-USB cord to the TP4056 charging board, connect the other end to any 5V cell phone charger or even my laptop, then wait. And watch. I don't see flames, so that's good. A couple hours later and the little red LED switches to green. My voltmeter shows me that the cell is at 4.20 V. Excellent, it's working so far.

Now I'll add the discharging board, which is my DC-DC converter. I simply solder the red positive wire from the battery to the "In+" pad on the discharging board followed by the negative black wire from the battery to the "In-" pad on the discharging board. I can also just add a wire from the pads on the TP4056 charging board too, since I connected them in the previous step.

A quick test with a multimeter shows me that the USB port is outputting 5 V, which means that it should be working. Now for the final test. I'll plug my phone's USB charging cable into the USB port on my Franken-charger and the other end into my phone. My phone lights up and starts charging. It's ALIVE!

Now I can seal the whole thing up. I should probably wrap the pouch cell in a thin sheet of foam to protect it from the hard case if I were to drop it, then I can just hot glue it into the case. The two circuit boards can sit on top of the cell or in the end of the plastic case, and I'll cut a little space for their USB ports as well. I should probably cut a hole so that I can see the charging LED on the charging board too.

And that's it. Done. Finito. Terminado. Gamur.

12s RC lipo electric skateboard battery

Electric skateboards can run on many different voltages. They generally use RC airplane motors and controllers, so that's what we'll assume. And because we're already using RC parts, we might as well go with an RC lipo battery for simplicity. It will make it easy to build the battery without any special tools such as a spot welder. Remember though, these batteries can be dangerous so we'll take proper precautions to ensure that we build a safe battery.

12s is a common configuration for an electric skateboard. At 3.7V per cell, that works out to a nominal voltage of 44.4 V.

We are going to build a fairly lightweight battery because we aren't going that far on our skateboard. Let's say we want 5 miles (8 kilometers) of range. Doing a little research shows me that 20 Wh/mile (12.5 Wh/km) is a fairly common efficiency figure for electric skateboards. That means we'll need 100 Wh to travel 5 miles. At 44.4V, that means we need a 2.25 Ah battery.

Required capacity of the battery in amp hours = 100 Wh ÷ 44.4 V = 2.25 Ah

Checking on the online RC shop HobbyKing, I can find 6s RC lipo batteries that are rated for 2,200 mAh (2.2 Ah). That's close enough for my needs. If I wanted to be safe though, I could always choose a battery pack with just a slightly higher capacity.

I'll need two 6s RC lipo batteries in series to make a 12s battery. The 6s batteries already come with bullet connectors, so I'll grab a few more of those along with some 12 awg silicone wire for my connections.

I'll tape the two RC lipo batteries together and make a short double-ended male wire with my bullet connectors by soldering a male bullet connector onto each end of a short length of wire. I will then use it to connect the positive terminal of the first 6s battery to the negative terminal of the second 6s battery. Now I've got a 12s battery. Technically it's a 12s1p battery pack, though it could also be described as a 2s1p pack of 6s RC batteries. Same thing.

The two discharge wires from the RC lipo packs that I didn't use will become the main discharge wires for my 12s battery. Those will plug into the controller on my electric skateboard.

For charging, I'll need a balance charger. An iMAX B6 is an inexpensive balance charger than can handle up to 6s batteries. I could charge each battery one at a time, but that'd be annoying. Instead, I'll purchase an RC lipo "balance board" which allows me to charge both batteries at once. I just disconnect the two 6s batteries from my electric skateboard's speed controller and plug both the discharge wires and their balance wires into the balance board. Then I plug that balance board into the iMAX B6 balance charger and it will charge both my batteries at once. When it's done charging, I can rewire my batteries in series (carefully so that I don't make a wiring mistake) and then plug them back into the controller on my electric skateboard.

A quick note: I shouldn't ever let these batteries charge unsupervised. If I leave the house or office before they've finished charging, I'll disconnect them and wait until I return to finish charging. Safety first.

One last safety step. It's critical that I don't discharge my RC lipo packs too low. Doing so can be dangerous and damage the batteries. To prevent this, I'll buy two RC lipo low voltage alarms. When I ride my electric skateboard, I'll plug one directly into each balance balance connector on my RC lipo batteries. They each have a small screen that tells me the voltage of every cell and I can set the voltage that the alarm will sound at to let me know that I've reached my desired cutoff.

Technically, I can let it drop down to about 3.0 V but that'd be dangerously low. There's no reason to take that risk. I'll set my cutoff at 3.3 V to be safe. Now when I ride my electric skateboard, if I ever hear the alarm, I know it's time to stop riding. I can also use it as a simple 'fuel gage' by checking the voltage of my batteries to know how close they are to empty.

And that's it. I've built a simple 12s RC lipo battery and I've taken a few extra steps to make it as safe as possible. One thing to note: I probably won't get the complete 2.2 Ah out of my batteries since I'm not discharging them down to 0% state of charge (as that'd be dangerous). So it might be a good idea to start with batteries that have a slightly higher capacity than what I think I'll need so that I can still go the same distance without completely discharging the batteries.

10s (36 V) electric bicycle triangle battery

Most electric bicycles use batteries in the range of 24 - 48 V. I'll build a 36 V battery and I'll aim for a capacity of 10 Ah, as that's also quite common in the ebike industry.

I plan to put my battery in a triangle bag in the center of my bike frame, but I want to use a small triangle bag to not take up too much space. So I'll build a triangle shaped battery to most efficiently use the space. I could build a rectangular battery, which would be much simpler, but then I'd need a larger triangle bag and it would waste a lot of space in the corners of the bag.

I'll use 18650 li-ion battery cells. I know I have a 15 A controller so I need my battery to have a continuous discharge current rating of at least 15 A. To meet the current and capacity requirements, I'll choose the NCR18650GA cell made by Sanyo/Panasonic. It has a capacity of 3,500 mAh and a maximum continuous discharge current rating of 10 A. I can use three cells in parallel to make 10.5 Ah, and three cells in parallel will also give me up to 30 A of continuous discharge current ability. That's twice what I need, which is great. That means I'll be using the cells well within their rated limits, which is much healthier for the cells.

This results in a 10s3p pack, which means I'm going to use 30 cells. Time to fire up my favorite Photoshop knock-off software and design a battery layout. The diagram shows what I've come up with for Side 1 of the battery.

Example of a 10s3p triangle battery layout

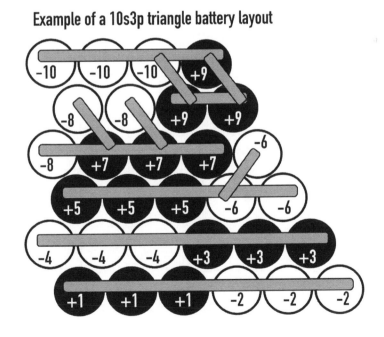

I also need a BMS. This time I'll head to my favorite AliExpress BMS source, which is a vendor known as "Greentime". They've got a 10s li-ion BMS rated for 30 A continuous discharge.

Again, that's twice the 15 A discharge current that I'll be using on my ebike, which means I will have a safety factor of 2 for the BMS as well. Excellent.

Now I can build my battery. I'll use a hobby level spot welder and join the cells using pure nickel strips of 0.15 mm thickness and 8 mm width. Since each of these pieces of nickel can carry around 5 A each, I'll need three pieces of nickel for each series connection to support my 15 A expected load. In some places I can weld a piece of nickel between each of the three cells in adjacent parallel groups (such as the +2 to -3 series connection, which is on the back side of the battery in the diagram). However, there are a few places where a single point connects two parallel groups (such as the +1 to -2 series connection). In those places, I'll need to use three layers of nickel.

To keep this battery small, I'll use hot glue to join the cells together instead of snap-together cell connectors. I'll start by gluing up the first two rows of 6 cells. Then I'll weld a single piece of nickel across the negative terminals of the three cells in the first parallel group on side 1. That will be my negative terminal for the finished pack.

Then I'll turn the battery over to side 2 and weld a long piece of nickel across the positive terminals of those same three cells in the first parallel group AND the negative terminals of the three cells in the second parallel group. That single piece of nickel covering all six cells just made a parallel connection in the first parallel group, a parallel connection in the second parallel group, and a series connection between the first and second parallel groups. But I need more nickel for this series connection between groups 1 and 2 because there is only a single point of connection. So I'll add two more strips of nickel: the first will be long enough to cover the inner four cells (two in each group) and the second will cover only the middle cells (one in each group). That gives me three strips of nickel which is enough nickel at every point in the series connection to handle the maximum amount of current that the battery will experience based on my 15 A load.

Now I'll turn the battery back over to side 1. Here I'll make the series connections between the positive terminals of group 2 and the negative terminals of group 3. The order doesn't really matter. I'll do the series connections first since they are more critical and I'd like to have the nickel in those connections welded directly to the cell terminals. Here I can weld a single piece of nickel between each of the three pairs of cells. That means I won't need to do any of that pyramid layering business. Nice and simple. After I do the three series connections, I can go back and add a parallel connection to the positive terminals of group 2 and to the negative terminals of group 3.

I'll flip the battery back over to side 2 and continue the connections in the same way. For the next connection between the positive terminals of group 3 and the negative terminals of group 4 I'll need to do the pyramid layering again since I have only one point of connection. Then I'll glue on one more row at a time as I continue to make connections all the way to the tenth parallel group.

Lastly I need to add my BMS. I'll solder the negative discharge and charge wires to the BMS board and then to the nickel strips on my battery's negative terminal (the negative terminal of parallel group 1). I'll solder the wires on the nickel in between the cells to make sure I don't heat up those cells too much. I could have also soldered the wire to the nickel first, and then spot welded the nickel, but that doesn't leave you a lot of room for spot welds if your wire is very thick. Then I'll solder on my positive discharge and charge wires to the nickel strip on the positive terminals of the tenth parallel group. Lastly I'll solder on the 10 thin balance wires.

To finish my battery, I'll wrap it in a layer of 2 mm EVA craft foam to provide a small amount of shock absorption and then use a few sheets of large diameter heat shrink to seal the battery. The triangle shape makes it a bit odd, but playing around with a few sizes of heat shrink will allow me to seal the whole thing from a few different angles.

Lastly, I'll need a charger. The same vendor on AliExpress that I got my BMS from, Greentime, also sells chargers. I'll select a 42.2 V charger, as that's the appropriate voltage for my 10s li-ion battery. An aluminum shell charger rated for 2 A charging will do nicely. That should give me about 5-6 hours for a full charge. Plenty of time for my needs. If I wanted a faster charger though, I can choose anything up to 4.5 A, as my cells are rated for around 1.5 A each for charging current. I don't want to push them too hard though, so 2 A total is fine for me.

And that's it! That's all I need to build a triangle battery for an electric bicycle.

38s (120 V) LiFePO$_4$ prismatic cell electric car battery

Now it's time for a big project. Let's build a battery for an electric car!

For cells, I think I'll go with a prismatic format. That way I don't need to worry too much about a protective case - many big prismatic cells for EV use already come in rigid plastic cases.

There are a number of good options in the LiFePO$_4$ chemistry, and they are both safer and longer lasting, so I'll choose that chemistry.

Now I need to calculate my needs. Let's say I know I'll be using a 120 V DC motor and a 200 A electronic speed controller. That means my voltage is decided for me, I need around 120 volts to power my controller and motor. For LiFePO$_4$ cells with a nominal rating of 3.2 V per cell, that means I'll need around 38 cells.

But how much capacity per cell will I need? Well, let's say that I want to travel around 55 miles on a charge. And I've done my research, so I know that compared to others with similar setups, my system should get an efficiency of around 200 Wh/mi. Now I can calculate my capacity needs, first in watt hours.

Required capacity of battery in watt hours = 200 Wh/mi × 55 miles = 11,000 Wh

So I need battery with a capacity of 11,000 Wh. Since I know the voltage I'll be using, I can now calculate the Ah I'll need for each cell.

Required capacity of battery in amp hours = 110,000 Wh ÷ 120 V = 91.7 Ah

That means I need to choose battery cells that are at least 91.7 Ah. Searching around, I find multiple sources for 100 Ah LiFePO$_4$ prismatic cells. I buy 38 and I'm nearly ready to go. I just need a BMS. I search around and find a 38s BMS that is specifically made for LiFePO$_4$ cells and can handle over 200 A of current. Now I'm ready to start planning my layout.

My battery cells are 2.5 inches (6 cm) wide, meaning that if I lined up all 38 cells, my battery would be nearly 8 feet (240 cm) wide! That's not going to fit in my trunk. Instead, I'll just make three rows of cells, which will have a footprint of about 2.5 ft × 1.5 ft (76 cm × 42 cm). That's more reasonable.

My prismatic cells have convenient screw terminals for connections. I decide that I'll use copper busbars to assemble my battery. I can either buy the busbars online or just buy some copper bar stock and drill the holes myself. I'll want to check with a busbar chart to make sure that the size of my busbars (and type of copper) are sufficient to carry the 200 A maximum current that my controller is expected to draw.

Because I'm using 100 Ah battery cells, I don't need to make any parallel connections. Instead, I can simply wire all of my cells in series as shown in the diagram.

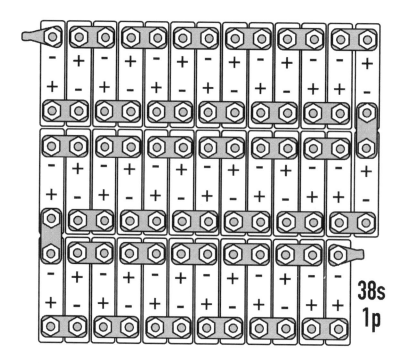

Lastly, I'll need to connect my BMS. It connects just like in my electric bicycle battery example above, but because I'm using copper bus bars I can just add a ring connected to the wires and I don't have to solder anything.

For a charger, I need around 140 V to fully charge my LiFePO$_4$ cells. Unfortunately, a charger that size isn't going to be cheap, but a company named Zivan provides me with a 140 V and 18 A charger that should charge my battery from empty to full in around 7 hours.

LiFePO$_4$ chargers are very commonly available for increments of four cells on the lower voltage ranges, such as 24 V, 36 V and 48 V (8s, 12s and 16s). One neat trick to save money on a high voltage LiFePO4 charger is to use a multiple of four cells, such as 36 cells, and then you could just use three 36 V LiFePO$_4$ chargers connect to your batteries in three subgroups. It's a little more work as you'll have to use multiple BMSs to make sub-batteries, but those lower voltage chargers are much easier to find.

14s (52 V) 18650 home energy storage battery

This type of battery is mostly known as a DIY powerwall and is perfect for supplementing your home energy generation project or for building an off-grid system. If you have solar panels, a wind turbine or any other form of energy generation on site, a DIY powerwall can store that energy to be used at a more convenient time, or sell it back to the grid during peak energy demand. Some people even use home energy storage batteries to "buy" electricity at night when its cheap and then use that electricity straight from their batteries during the day when it would cost more to get it from the grid.

There is a big DIY powerwall community online that spans a number of websites and forums. The vast majority of DIY powerwall builders use salvaged 18650 cells, so that's what we'll do in this example.

I'll assume that I want to use a 48 V inverter for my energy use, so that means I need a 13s or 14s li-ion battery. The best option is 14s, as the low voltage cutoff of around 42 V is more in line with most commercial 48 V inverters.

I can calculate my battery needs by determining how much energy I want to use each day. Let's say that I have a goal of storing at least 5,000 watt hours. For a 52 V li-ion battery, that means I'll need around 96 Ah. Let's just round up to an even 100 Ah.

Required capacity of battery in amp hours = 5,000 Wh ÷ 52 V = 96.2 Ah

I stated that I want to use salvaged cells. Assuming I can find a pile of approximately 2 Ah 18650 cells, that means I'll need around 50 cells per parallel group to reach my goal of about 100 Ah. Notice that everything is "about" and "around" and "approximately". Such is life when you're using mystery salvaged cells.

Ok, time to go dumpster diving. Or more likely, time to visit the local battery recycling center. After buying about 150 lbs (70 kg) of recycled battery packs, I can crack them all open and pull apart the battery cells. Now I've probably got at least a thousand or so 18650 cells. I can use a 4-cell 18650 tester, or better yet an army of those testers, to check the capacity of all of those cells. This will take a few weeks depending on how many cells I can test at once.

I ultimately end up with 700 cells that are all fairly close in capacity and somewhat above 2 Ah. Some might be 2,050 mAh, some might be 2,110 mAh, but they are all fairly close. Generally we would want all cells in a parallel group to be exactly the same capacity, but we'll take the best we can get when working with salvaged cells.

I'll prepare the cells by sanding their terminals smooth to remove any spot welded nickel shards that try to hang on there.

Now I'll divide them up into groups of 50 and try to keep them as close to the same capacity as possible. Then I'll construct parallel modules of 50 cells using snap-together block connectors.

Ideally these 50 cells in each module would be spot welded together. However, the DIY powerwall community usually solders these, so that's what I'll do in order to demonstrate that method. They also usually use cell level fuses, so I'll do that too.

I cut a pile of legs off of ¼ W resistors to use as the cell fuses, then solder them to the cell terminals using a thick soldering iron tip to hold in the heat. I might have to use flux as well, depending on the cell terminals. I contact the soldering iron tip to the end of each cell for as little time as possible so that I don't transfer too much heat to the cell.

The other ends of the fuses are soldered to one or more copper busbars. Those will make the positive and negative terminals for the entire module.

Once my parallel groups are assembled, I'll wire them in series by connecting each of the busbars to the next parallel group until I've built an entire 14s50p battery. Copper wire with ring connectors on bolts passing through the busbars would be a good way to wire them in series.

Lastly, I'll need to add a balance connector with a balance wire that connects to the positive terminal of all 14 parallel groups. Most DIY powerwall folks don't use a BMS but rather use a balance charger. A big 14s balance charger isn't cheap, but it's what I need for this battery to keep all of those mystery cells happy and working for as long as possible.

Ideally I'll use this pack at a low C rate; 0.5 C would be a good upper limit. That means each cell won't be delivering more than 1 A, which should keep them happy and healthy. At 52 V, a maximum current of 50 A would give me around 2,000 W of maximum continuous power. That's enough for my off-grid doomsday cabin. Besides, I wouldn't want to run too many devices at once and risk drawing attention from the zombies.

4s (14.8 V) 18650 RC drone or FPV battery

A great use for DIY lithium batteries is for RC drones and aircraft. Most of these aircraft use RC lipo batteries, but we can actually build a lighter 18650 battery that will give more flight time with less weight. For people in the first person view (FPV) flying community, fight time is key.

Let's say I've got an RC FPV airplane that operates on 4s (14.8 V). And let's say that I generally use an 8 Ah RC lipo pack that weighs around 415 grams. We can build something better than that!

Let's start with our current requirement. Let's say my airplane draws a maximum of 30 A but generally cruises at around 10 A. If I use the same NCR18650GA cells from the electric bicycle battery example earlier in this chapter, I could use three cells in series to get a maximum continuous current rating of 30 A. That's the extreme limit of the cells and exactly matches the peak current our airplane uses. That would be a bit close for comfort if that was the airplane's continuous current, but my airplane usually cruises at around 10 A. So my battery will only see bursts of 30 A when the plane is climbing aggressively. That's fine by me and within the limits of the battery cells.

Three cells in parallel would give me 10.5 Ah, which will provide even more flight time than my old 8 Ah battery. I'll need a 4s3p arrangement for a total of 12 cells.

I'll just hot glue the cells together to save on weight and space. Then I'll spot weld them using the same method I used in the electric bicycle battery example. However, it will be much easier this time as I've got straight rows of parallel groups.

If I'm using nickel strip that can handle 5 A per strip, then I'll need six pieces to support the 30 A maximum current that I expect the battery to supply. Since I have three cells in each parallel group, I'll use two layers of nickel per series connection between each pair of cells. That will give me six pieces of nickel strip for every series connection.

I won't use a BMS because I want to save on weight, so I'll solder my discharge wires directly to the first and last cell group and add a 4s balance connector. That means I'll have to use a balance charger, such as an iMAX B6, to safely charge and balance my battery.

Lastly, I'll use a piece of heat shrink to cover the whole battery so the exposed terminals can't be shorted on anything. This results in a battery that has 2 more Ah than my original RC lipo and weighs about 200 g less. That weight savings means even more flight time. Time to go cruise the skies!

Conclusion

I hope you found this book both informative and useful. Lithium batteries will likely be powering our devices for decades into the future. Learning how to build your own batteries opens the door to a whole new world of custom powered devices. From vehicles to on-site energy storage to games and even wearables, making your own custom lithium batteries enables you to build and create on a whole new level.

With that in mind, it is important to always make safety a priority when working with lithium batteries. With the good comes the bad. Lithium batteries drive technology forward but have also caused death and disasters when engineers and designers cut corners or ignore standard safety practices.

As Ben Parker said, "with great power comes great responsibility". You've been given the knowledge, and now you must use it responsibly. Accidents due to improper lithium battery use leave a lasting mark on the entire industry. Please help by being a force of good moving the industry forward in a safe and responsible way.

Acknowledgements

The idea for this book was born out of a conversation with a friend of mine. Thank you, Elie Fuhrman, for sparking my creativity. I of course need to thank my wife as well for putting up with me, not just while I wrote this book, but also for years of finding half completed batteries laying all around the house with "danger - don't touch!" signs on them. My parents taught me to follow my dreams and not stop until I could say that I'd either accomplished my goal or proudly gave it everything I had. I appreciate everything they've done for me and gratefully acknowledge that if I'm a decent person then it's due to their excellent parenting. My dad started me on my journey to becoming a Maker, and when I went off to college, my mentor Andy Holmes picked it up from there. I think I learned more from working in Andy's shop than I did from my entire engineering degree. And I learned a LOT from my engineering degree. I owe a big thanks to Damian Rene who traveled down the lithium battery-building road with me as we simultaneously learned more about battery building. Being able to bounce new ideas and discoveries off of each other was an incredible resource. Lastly, I'd be remiss if I didn't thank Max Pless and Thorin Tobiassen, who were there during a pivotal time and helped me get into ebikes. This ultimately started me on a wild ride (literally and figuratively) that in part led to where I am today. Thanks guys.

About the author

Micah Toll is a mechanical engineer and entrepreneur with nearly a decade of experience in the electric bicycle industry. Micah's book *The Ultimate DIY Ebike Guide* has sold thousands of copies all over the world. He has built nearly every type of custom lithium battery for consignment and for his personal projects. Micah believes in the principle of "Don't buy what you can make" and promotes a Maker lifestyle based on handiness, resourcefulness and skill collecting. He currently lives in Tel Aviv with his beautiful wife Sapir and his dog Seven.

References

The information in this book comes from a combination of the knowledge I've gained over years of experience and diligent research from academic and industry literature. I hope these sources will be of as much use to you as they have been to me.

al, Qingsong Wang et. "Thermal runaway caused fire and explosion of lithium ion battery." Journal of Power Sources 208 (2012): 210-224.

al, Riza Kizilel et. "An alternative cooling system to enhance the safety of Li-ion battery packs." Journal of Power Sources 194.2 (2009): 1105-1112.

Buchmann, Isidor. BU-205: Types of Lithium-ion. 30 November 2016. 31 January 2017 <http://batteryuniversity.com/learn/article/types_of_lithium_ion>.

—. BU-410: Charging at High and Low Temperatures. 2 April 2016. 22 January 2017 <http://batteryuniversity.com/learn/article/charging_at_high_and_low_temperatures>.

Dahn, Jeff. "Why Do Li-ion Batteries Die and Can They Be Immortal?" WIN Seminar. Waterloo Institute for Nanotechnology, Waterloo. 30 July 2013.

Golubkov, Andrey W. et al. "Thermal-runaway experiments on consumer Li-ion batteries with metal-oxide and olivin-type cathodes." Royal Society of Chemistry 7 (2013).

Marom, Rotem, et al. "A review of advanced and practical lithium battery materials." Journal of Materials Chemistry 27 (2011).

Search for the Super Battery. Perf. David Pogue. Prod. Chris Schmidt. Public Broadcasting Service, 2017.

Svoboda, James A. Introduction to Electric Circuits. Vol. 9th Edition. New York: WileyPLUS, 2013.

van Schalkwijk, Walter A and Bruno Scrosati. Advances in Lithium-Ion Batteries. New York: Kluwer Academic Publishers, 2002.

Various Authors. 2007-2017. 2010-2017 <www.EndlessSphere.com/forums>.

Wu, Yuping. Lithium-Ion Batteries: Fundamentals and Applications. Boca Raton: CRC Press, 2015.

Yuan, Xianxia, Hansan Liu and Jiujun Zhang. Lithium-Ion Batteries: Advanced Materials and Technologies. Boca Raton: CRC Press, 2011.

Feel free to use these templates for planning out your own cylindrical battery packs

Simply shade in cells to indicate polarity and then draw your connections

Feel free to use these templates for planning out your own cylindrical battery packs
Simply shade in cells to indicate polarity and then draw your connections

Index